Severe Weather Warnings: An Interdisciplinary Approach

William Donner • Havidan Rodriguez •
Jerald Brotzge • Walter R. Diaz

Severe Weather Warnings: An Interdisciplinary Approach

Springer

William Donner
Department of Sociology
University of Texas Rio Grande Valley
Edinburg, TX, USA

Havidan Rodriguez
University at Albany, State University of New York
Albany, NY, USA

Jerald Brotzge
Ogden College of Science & Engineering
Western Kentucky University
Bowling Green, KY, USA

Walter R. Diaz
College of Liberal Arts
University of Texas Rio Grande Valley
Edinburgh, TX, USA

ISBN 978-3-031-05033-6 ISBN 978-3-031-05031-2 (eBook)
https://doi.org/10.1007/978-3-031-05031-2

This Springer imprint is published by the registered company Springer Nature Switzerland AG
The registered company address is: Gewerbestrasse 11, 6330 Cham, Switzerland

Engineering is not merely knowing and being knowledgeable, like a walking encyclopedia; engineering is not merely analysis; engineering is not merely the possession of the capacity to get elegant solutions to non-existent engineering problems; engineering is practicing the art of the organising forces of technological change... Engineers operate at the interface between science and society.
Gordon Stanley Brown

The ideal engineer is a composite. He is not a scientist, he is not a mathematician, he is not a sociologist or a writer. But he may use the knowledge and techniques of any or all of these disciplines in solving problems.
Nathan Daugherty

We define an engineered system as a combination of components that work in synergy to collectively perform a useful function.
National Science Foundation

In Memorium

We dedicate this volume to Dennis Mileti, who, along with John Sorenson, Michael K. Lindell, Marjorie Greene, and Ronald Perry, built through decades of thought and research the scholarly foundations on which rests our grasp of warnings. Dr. Mileti's insights are so fundamental to our present understanding of warnings that we can say with confidence that this volume—and many other academic works on warnings, for that matter—would simply not exist had it not been for his insights into the nature and character of the warning process. His work changed in basic ways how we, as students of disaster, regard not only the warning process, but risk in general. In highlighting the social, cultural, and psychological factors governing the warning process, he worked to dispel the pervasive and dangerous myth that fear and instinct—premised on the far-too-simple "stimulus-response" model of human psychology—were the sole factors shaping how humans respond to warnings and risk in general. It truly is impossible to overstate his contribution to our understanding of risk, warnings, and the field of Sociology of Disaster. With Dr. Mileti's passing, the field has truly lost a great mind and one of its most innovative thinkers. After finishing the book, this will be transparently clear to the reader, as his thought forms the foundation for the general theories and ideas one encounters in the following pages. Dennis will be greatly missed.

Preface

If pressed to summarize and describe the central argument of this work, we might reply accordingly: that the engineered system, from the humble, primitive pulley and lever to the modern, computer-driven numerical weather prediction model, stands inscribed with testable assumptions about patterns of human thought and behavior. Above all else, the interdisciplinary perspective with which this book is centrally concerned takes for granted that applied scientific questions naturally and implicitly express hypotheses regarding the way people act, react, think, behave, judge, believe, and anticipate. Systems or technologies so designed without regard for the user or beneficiary will therefore result in little more than technical innovation at the expense of community health, happiness, and safety, the fulfillment of which is the ultimate concern of engineering and other applied sciences. We might go as far as to say, at the risk of being accused provocative, that one cannot successfully conduct applied scientific research without the aid of social science—that is, if one defines success in the domain of science as bringing forth some ultimately beneficial outcome for society. A *useful function*, as it were. Today we enjoy the benefits of highly-sophisticated radar systems, talented forecasters and emergency managers, and the advanced infrastructure to broadcast warnings as quickly and as accurately as possible. Unimaginable innovation has taken place in the fields of meteorology, radar engineering, and warning communications. Nevertheless, as any emergency manager from a tornado-prone region can attest, *many simply do not take warnings seriously*, leading to hesitation or inaction, the effect of which is to cancel out whatever benefit such talent and technology may have brought forth. Thus, science and engineering, on the one hand, and the social sciences on the other, are, if viewed in the context of improving society, altogether inseparable efforts.

This leads to what is perhaps the central question of this volume: what is a "better" warning? Is it a warning that simply arrives earlier and with greater accuracy, or is it a warning that the public understands, believes, and, as it is hoped, responds to? The social science component provides that the latter is accounted for. Without psychologists, how are we to know whether new radar data will overwhelm and overtax emergency management decision making? Without sociologists, how are we to understand the cultural and structural limitations the public faces when responding to warnings? Without economists, how are we to know the economic feasibility of the radar system for institutions and the community? These questions are as important to warning systems as questions related to radars, algorithms, and weather prediction models.

Turning to the broader relevance of this work, the volume intends to cover as comprehensively as possible a broad range of topics, themes, ideas, and theories for what is perhaps an equally broad range of audiences. For meteorologists, forecasters, and emergency managers, we hope the forthcoming pages offer some guidance regarding how institutions and the public respond to hydrometeorological threats. Interdisciplinary teams—composed of natural and social scientists—at the earliest stages of project development may find a "how to" guide for organizing an interdisciplinary project, as well as strategies for maintaining a cooperative focus across the many challenges that may arise in the oftentimes complex effort of collaboration across the disciplines. Perhaps the volume will also spark a broader debate on the nature of interdisciplinarity and science—we hope so, for we regard attempts to address the question as in need of greater attention moving forward. It should perhaps also be noted that even though what follows deals specifically with interdisciplinary severe weather warnings, the book should nevertheless remain general enough in scope to assist STEM researchers and social scientists in developing collaborative partnerships on nearly any topic—that is, as long as the general question is of an applied character and addresses risk in some form or another. Though this is perhaps too far a digression.

It might be asked what were the motivations for writing a book on interdisciplinary severe weather research. First, the authors of the following volume share between them decades of experience conducting collaborative research and felt it was perhaps due time to synthesize what both they and the scientific community have learned on the issue of interdisciplinary severe weather, as well as what is known about the general theory of interdisciplinary research. Second, because there is no grand, overarching, or accepted theory of interdisciplinary integration. As it were, there is currently no "road map" for interdisciplinary research. In later chapters, we provide guidance on the development of an interdisciplinary project, outlining new roles, common needs, and common challenges of an organized interdisciplinary research effort.

But none of this is new, of course. For to advance, as we have, that applied scientific and engineering questions reflect questions of human behavior is to simply take to a logical conclusion the quintessential axiom of the Sociology of Disasters: that disasters are not fundamentally natural but rather *social* in character as Dynes and Quarantelli famously observed. Or, as put by Gilbert White, one of the most influential geographers of the twentieth century: "Floods are an act of God, but flood losses are largely an act of man." These ideas and the decades of empirical research that followed in which these ideas were tested served to overturn the notion that tornados, floods, hurricanes, earthquakes, and chemical spills bring forth risks by simply the virtue of their occurrence. In the present day, one does not have to look far for proof of this idea. Outside of severe weather, the unthinkable human toll associated with the refusal to wear masks or vaccinate oneself against COVID or to adopt measures to reverse climate change, both of which ultimately boil down to factors beyond the virus and factors beyond the greenhouse effect, relate fundamentally to the decisions made by people, the social structures by which they are bound, their expectations and management of risks, their conduct, and their values and beliefs. In this book, we take these principles and ideas as central to the analysis and application of interdisciplinary warning systems around which all discussions and arguments shall revolve.

In closing, we argue by implication that we cannot naively export engineered systems—warning-related or otherwise—into social systems. The engineered and social system must be first made compatible in some way such that the former is adapted to the conditions of the latter. Ultimately, with this general philosophy in mind, we hope the following volume brings forth a debate on the character of warning systems, the fundamental nature of engineered systems, and the relationship between the natural sciences and engineering fields, on the one hand, and, on the other, social science. It is perhaps the overall intention of this book to discover and elucidate a model for finding that imperative-yet-elusive "useful function" so described by NSF in a society of changing size, complexity, and character against which stands risks of an unprecedented scope and quality.

Edinburg, TX — William Donner
Albany, NY — Havidan Rodriguez
Bowling Green, KY — Jerald Brotzge
Edinburg, TX — Walter R. Diaz

Acknowledgements

The work was made possible in part through the generous support AXA Research Fund, as well as the efforts of our research assistants Santosh Mukingal and Amanda Mercado whose dedication to the project through long hours of research, reading, and revision made this volume possible. We also thank earlier support through the National Science Foundation (NSF) and the Engineering Research Center for the Collaborative Adaptive Sensing of the Atmosphere (ERC-CASA) in providing the resources and collaborations upon which many of the ideas and arguments in this volume were made possible.

Contents

Chapter 1
Introduction

The 2011 Joplin, Missouri, tornado disaster would claim 158 lives and injure 1200 more in what would be one of the most devastating and destructive severe weather outbreaks in recent history. The Joplin example is relevant to the study of interdisciplinary warnings due to the disproportionately high fatality and injury rates despite 19 min of lead time (Paul and Stimers 2014), an interval far exceeding the national average of 13 min lead time witnessed generally during the period, which, all else equal, should have resulted in a loss of life far less than recorded. Yet this was not so. How are we to understand and make sense of such contradictions that despite nearly a half hour in which to seek shelter from the storm such extraordinary loss could still occur? While the severity and complexity of the storms offer some explanation, other answers might be found by looking at the victims themselves, or, more precisely, how the victims regarded warnings at the time of the storm. Some insight is revealed, for instance, in the fact that Joplin residents were observed to ignore warnings (*St. Louis Dispatch* 2011). In other cases, the need for confirmation delayed shelter-seeking. Sirens were reportedly disregarded or greeted with skepticism (*Reuters* 2011). Such observations raise important questions. What are the circumstances under which warnings are heard, are heeded and what are the circumstances under which they are not? Why was the need for confirmation so widespread and prevalent? Did confirmation lead to delays in sheltering, and, if so, did these delays contribute to mortality? Were there circumstances when some sheltered, but it wasn't enough? The storm severity may have resulted in a high number of deaths, but in what ways did these factors contribute to the large number of injuries, as well?

The need for a greater focus on these questions emerges from recent statistics on mortality and severe weather. Despite a net decrease in tornadic fatalities since the 1950s (Brooks 2009), Stokoe (2016) observed that the decline diminished over the past three decades, a pattern similarly witnessed nearly a decade earlier among scholars (Ashley et al. 2008). This is perhaps because science and technology alone, however advanced at predicting and detecting the weather, may only achieve a reduction in fatalities to a certain point beyond which other factors are responsible for shaping severe weather mortality. Thus, designing warnings that provide longer

W. Donner et al., *Severe Weather Warnings: An Interdisciplinary Approach*,
https://doi.org/10.1007/978-3-031-05031-2_1

lead times, more accurate storm predictions, and fewer false alarms may be a necessary but insufficient condition to improve public response. "There are limits," observes Folger (2011: 20) in a congressional report, "to the number of potential lives saved by improvements in forecasting ability and warning systems, and that social, behavioral, and demographic factors may play an increasingly important role in tornado-related fatalities." This is not to diminish the role of technology. Vast developments in infrastructure and systems over the past century have led to dramatic improvements in the speed and accuracy of warnings. But it may be time to acknowledge that technical improvements in warning systems may now have reached a point of diminishing returns on public safety.

Warning: A Theoretical Definition

To begin this section, we present a paradox. It is indeed a curious thing that in the middle of a remote country road, with no perceptible cross traffic and without a police officer in sight, and accordingly no threat of consequences or accountability, few overcome the impulse to stop at a red light or stop sign if encountered. Under such circumstances, the possibility of an accident or fine is remote—not a car in sight, no risk of collision, not police officer in sight—yet few would even dare think to simply drive through the intersection in heedless defiance of a clear signal to stop. Indeed, most would likely bring their vehicle to a stop with hardly little reflection on the matter as is customary. Compare this scenario with tornado warnings. Tornado sirens are routinely *ignored*, even in the presence of darkening skies, the pleas of broadcasters and emergency managers, and, in other cases, even in the incontestable presence of tornadoes themselves. At face value, this seems puzzling to the casual observer: *why stop at a stoplight on a desolate country road when there is no perceptible risk, but ignore a tornado warning when the potential consequences are laid bare to the observer?* Only if we redirect our attention away from stoplights and tornado alarms as specific technologies *per se* to the general, defining character of a warning might we begin to understand this paradox. To proceed scientifically, we need to define a warning in its most general and abstract terms, which is to say we must first identify the common and essential features of a warning, eliminating those features which are merely incidental.

Before pursuing such a theoretical definition, it is perhaps useful to define what a warning *is not*. A warning is not merely the din of a fire alarm, ominous text on a cigarette package, the blare of a tornado siren, or a stoplight. It is not the wires, buttons, and electrical circuits, nor is it the loud sounds or flashing lights that makes a warning truly a warning. While these are important warning devices and methods, they do not comprise warnings in and of themselves, but only the incidental features of warning systems. In order to understand warnings scientifically and theoretically, we need a broader, more general definition, one that identifies the most universal qualities of the warning process common across different warning systems (stoplights, sirens, alarms) and scenarios (e.g., traffic safety, tornado response, fire evacuation). We define warnings accordingly: in the most universal sense, all warning scenarios necessarily involve a stimulus intended to (a) alter perceptions of risk, and (b) motivate action. Warnings of any source attempt to convince that the level of risk one faces is greater than the normal, everyday threshold and to further persuade that some action is necessary to return risks to that threshold at the very least. Defined in such a way, the psychological and behavioral elements of risk, such as biases and heuristics (Tversky and Kahneman 1974), are allowed to take a more central role in the scientific study of warning, thereby moving away from insufficient "stimulus-response" models of the warning process.

With this theoretical definition in mind, why the public tends to ignore warnings begins to make greater sense. It allows us to understand, for instance, why written cigarette warnings remain so notoriously ineffective (Monárrez-Espino et al. 2014)—namely, because the decision to smoke is fundamentally shaped by forces well beyond the mere text on a cigarette package. In terms of the cigarette warning itself, this may be because simple, descriptive text on cigarette packaging conveys little sense of immediacy or urgency to the would-be smoker and therefore evokes a lukewarm response to the risks of cancer or emphysema or any other health outcome to which the reader is alerted. Research that went into the development of the famous CDC *Tips from Former Smokers* campaign, for instance, suggested that anti-smoking messages, if to be effective, must evoke clear images of the impacts of smoking on the lives of smokers and their families, the absence of which may account for the failure of simple text warnings to deter smoking. What is more, oftentimes the decision to smoke remains dependent on broader social change. For instance, cultural shifts in society, which increasingly frame smoking in moral terms, have seen smoking behavior gradually stigmatized, a factor responsible for reduced rates of smoking (Bayer 2008; Chapple et al. 2004). Higher tax rates (Bader et al. 2011) and changes in media portrayals of smoking (Carson et al. 2005), as well, also discourage potential smokers. Peer influence is so well established a cause that there is little need to evoke the literature. The point is clear: these are social forces unrelated to the warning text on a cigarette package, yet at the same time shape the decision to start or continue smoking. The central lesson is that warning decisions cannot be reduced to the warning itself. A warning of any kind or form is a necessary yet insufficient factor in the response process. Why should this be any different in the context of tornado warnings, or any other scenario in which one is alerted to the presence of risks?

This leads to the inevitable conclusion that warning technology—be it a cigarette warning so described or any other warning, for that matter—in and of itself cannot alone motivate response as response is controlled by a dynamic, multidimensional, and complex set of social and psychological processes. In other words, warning response cannot be simply reduced to cigarette messages, or stoplights, or tornado warnings, or fire alarms *per se*. This forces us to think more broadly about warning systems, for, under such a perspective, ignoring warnings cannot be reduced in the final analysis to a simple cavalier attitude towards risk or a frivolous interpretation of warning messages, but remains the product of something far deeper and far more complicated. Indeed, it is tempting to simply label those who disregard warnings as complacent, irresponsible, cynical, or chronically indecisive, but this does not explain why they have become complacent or cynical or irresolute in the face of warnings in the first place. These are far too uncomplicated explanations, for something apart from the warning itself had to bring about these attitudes in the first place. We might then ask: why are certain warnings responded to, and others not?

Why Are Some Hazard Warnings Heeded and Others Not Heeded?

Before delving into the history of warning systems, here we offer a further illustration of the social nature of the warning process through an extended comparison of two disaster-related warnings: that of the difference between fire alarms and severe weather alerts. What makes the comparison instructive and worth examining in greater detail is that it reveals the unmistakable social character of warnings. Once again, the origins of the difference in response likelihood must, according to the definition presented, lie beyond the properties of the technologies themselves (both, for instance, involve loud sounds), so answers must be sought elsewhere: in the social, cultural, structural, geographic, and psychological contexts within which either fire alarms or tornado sirens are received.

First, consider differences in rates of reception between fire alarms and tornado warnings. Few can ignore a fire alarm. Nearly everyone in a building will hear a fire alarm if activated and functioning as intended. In contrast, tornado warnings are far more likely to go unnoticed—poor internet connections, sleeping, walls obstructing the sound of outdoor sirens, isolation from community networks through which warning information circulates, and so on, may limit a widespread awareness of the alert within the community. All else equal, fire alarms are simply more likely to be received in comparison to tornado warnings. Naturally, if one is more likely to hear a warning, whatever form that warning may take, one is more likely to respond.

But reception is only one of many factors in varying response rates. There are, furthermore, few if any social or economic costs to heeding a fire alarm (indeed, there may be benefits: leaving the building may offer respite from a boring workday!) and there is great pressure to conform when others are witnessed evacuating. Building occupants hearing a fire alarm require few, if any, environmental cues to confirm the need to evacuate. Flight often occurs without the need to see or smell smoke or witness a fire directly. One learns to heed fire alarms "just to be safe." Social cues, in the form of others leaving the building, often accompany a fire alarm, thereby encouraging others to leave as well. Witnessing others evacuate, even in a calm, orderly manner, may be regarded as reason enough to exit a building. At the same time, there is a certain degree of social authority invested in fire alarms, such that people feel compelled to react. No parallel culture exists that dictates immediate tornado response. (The same can be said for other weather, earthquake, volcano, and tsunami alerts.) As a result, most tornado warnings, be they in the form of a siren, NWS warning, or televised alert, often call for confirmation before action is taken (Lindell et al. 2016), a behavior notably less common in the context of fire alarms. Additionally, unlike house fires, tornado warnings evoke fewer *visible* social cues given that people are more widely distributed across neighborhoods and cannot readily perceive whether neighbors have taken any action to protect themselves. The absence of such cues further serves to reinforce a sense of normality. Thus, it is largely the context in which warnings take place, rather than the warning technologies or systems themselves, that shapes response. It is this warning context that illustrates the general definition and theory of warnings as presented earlier. The act of sheltering from severe weather, alternatively, comes at a greater perceived cost (should I take off work?) and in a context of greater ambiguity in the absence of social cues that would signal the need to take action.

Finally, at the level of psychology, the sound of a fire alarm almost immediately revokes the perception of normality and creates a sense of urgency, the sense that "something is not quite right." As a result, many feel compelled to evacuate. Tornado warnings, by comparison, produce a blunted psychological effect. Far from signaling a break from normality, tornado alerts are often swept into the currents and rhythms of a community's social life, oftentimes ignored in areas where severe weather and severe weather warnings are a common event. In cases where tornado warnings *do* manage to provoke some concern, response is rarely immediate, instead resulting in a process by which warning recipients weigh the costs against the benefits of sheltering. They ask: Is it worth my time to seek shelter? Can I take time off work? Should my children miss school? Is it worth sacrificing my leisure activities?[1] All in all, complex calculations of personal safety, nearly absent in the context of fire alarms, often favor inaction in the context of tornado warnings. Again, the example of fire alarms and tornado warnings further illustrates that warning response cannot be reduced to mere fear or instinct. While warnings must be clear and timely, this is simply not enough.

The Value and Limits of Traditional Warnings

In the absence of a formalized vocabulary on the matter, here we introduce the term "traditional" warnings to describe a warning system regarded in mainly technical terms. So conceived, traditional warning research focuses principally on improving the quantitative, technical elements of warning system design. This has resulted in important advances such as more accurate, higher resolution meteorological models to reduce storm prediction errors; optimizing remote sensing methodologies in

[1] During a somewhat recent tornado outbreak, a weatherman was berated via social media for interrupting the airing of a popular reality television show (see: https://www.esquire.com/entertainment/tv/a27613375/ohio-weatherman-bachelorette-tornado-warning/).

efforts to yield faster, more reliable radar data; and developing technologies that speed up communications. The logic of these efforts is fairly uncomplicated: if storm systems are identified more quickly and with less ambiguity, this should provide the public with more time to seek shelter, which, in turn, should minimize fatalities. At face value, few would question the *modus operandi* of traditional warning research, for who would want *less* time to shelter or *less* precise warnings? But it is necessary to ask: are such advances a benefit to public safety as intended, or do they simply exist as improvements to systems from a *purely technical standpoint*? This is perhaps the central question of interdisciplinary research. It is not about whether the public needs *more* or *less* time to respond, but rather the *correct* time to respond. It might not be whether a warning is *more* or *less* precise, but rather how the public defines precision and what level of precision is ideal to motivate understanding and response. In the end, the value of a warning system comes down to lives saved and injuries prevented, a goal only partially achieved through system innovation. As was seen in the case of Joplin, traditional warnings may have reached a point of diminishing return.

Accordingly, further reductions in fatalities and injuries may not be possible without an expansion of the traditional model of warnings. This is because, as was shown, warning response cannot be reduced to mere instinct or fear, which traditional warnings take as the primary motivating factors in the decision to heed and act upon warnings. Rather, much happens between the point of reception and the final act of response, resulting in delays or inaction that place people and communities at risk even if warnings are precise and timely. Decades of scholarship suggest that upon receiving a warning, recipients do not act immediately, but routinely pass through a series of progressive stages in which are raised questions regarding certainty and credibility [this process is discussed in detail in forthcoming chapters, readers may find the model elaborated fully in Mileti and Sorensen (1990) and Lindell and Perry (1992)]. Is the message received and understood? Is it credible and, if so, personal enough to arouse concern? Will recipients feel the need to confirm warning information? Do they understand what to do and what resources are available for protection? Finally, even if prepared to take action, are the necessary resources available to respond accordingly? A "no" to any of these questions—due to skepticism, misunderstanding, or lack of means, any of which may be perceived or real—at any stage in this process results in inaction or delay irrespective of the sophistication of the radar system or other tool from which the warning was produced. Adding further to this complexity, the progression of warning stages more often than not happens within a distinct social and cultural setting and therefore the process of evacuation and warning response is profoundly altered by social context (Donner 2007; Peacock et al. 1997). In other words, no two communities respond to a warning in the same way as no two communities enjoy the same advantages, face identical limitations, or observe similar cultural values and norms of interaction. It is for these reasons that traditional warnings cannot comprehensively and effectively address the process of public response across different populations.

Thus, although traditional warnings have led to great strides in reducing mortality, there remains a tremendous need for "connecting the last mile"; in other words, engaging the recipient and communicating the warning message effectively. Warning systems must be adapted to the observed and established realities of warning response, which occurs far beyond the confines of the immediate technical system. In this respect, interdisciplinary research, which involves collaboration and teamwork between the social and natural sciences offers one way through which to revise warning systems in a manner consistent with the realities of warning response among communities and community members.

Traditional vs. Interdisciplinary Warnings

How then might we draw an analytic distinction between *traditional warnings*, so described, and what might be called *interdisciplinary warnings*, the focus of this volume? To reiterate briefly, traditional warnings approach the design, research, and application of warning systems from a principally scientific and engineering perspective. Human factors, while considered, often are of secondary concern in the traditional perspective. Communities should *adapt to* the design of the warning system. Interdisciplinary warnings, on the other hand, retain a critical emphasis on the technical design of warning systems (for this naturally cannot be ignored), but also ask how psychological, social, and community factors such as culture and social structure shape how the system is used, and how this information should be incorporated into system design preemptively to maximize warning effectiveness. Accordingly, interdisciplinary warnings do not ask communities to adapt to pre-designed systems, but rather seek *to adapt* systems to the community. Interdisciplinary warning research places the needs and limitations of key stakeholders—forecasters, emergency managers, and the public—at the forefront rather than the periphery of the design.

As far as research is concerned, here the interdisciplinary effort tasks social science team members to provide evidence-based guidance and support to the scientists, engineers, and meteorologists developing the technical elements of the system. The process of collaboration is continuous and iterative, spanning most, if not all, of the design stages. The philosophy of interdisciplinary research naturally implies an overarching principle that all participants—both natural and social scientists, as well as key stakeholders and the public—must be involved at all stages of the process—from the earliest stages of conception through the end stages of system evaluation and adjustment. In the context of interdisciplinary warning research, ignoring the needs of stakeholders risks developing an irrelevant or at the very least inefficient system. A new radar system, for instance, might offer faster, higher resolution data and products, which, while advantageous in some regards, could simultaneously overburden decision makers, tax local computing resources, and exceed the training limits of forecasters and emergency managers. Without feedback from users and recipients, a warning system may be too complex or provide the wrong kinds of information or, alternatively, might provide *too much* information (even if theoretically useful), thus neutralizing any potential benefit of technical innovation. Ultimately, the interdisciplinary approach seeks to ensure that each end-user or stakeholder group ultimately finds use in or benefits from the system design. The integrated system must add value to the decision making and judgments of end-users if the systems are to be useful.

A Note on Human Factors and Ergonomics

Before exploring in-depth interdisciplinary ideas, it is first necessary to distinguish between human factors research (oftentimes used interchangeably with the term "ergonomics") and interdisciplinary research. Human factors forms only part of the interdisciplinary endeavor. The two areas overlap, but critical distinctions exist that should be highlighted.

Human factors research serves as an important element in the integration of engineering systems. While many definitions of human factors exist in a variety of different engineering fields, one useful and sufficiently general definition suggests that "human factors engineering focuses on how systems work in actual practice, with real—and fallible—human beings at the controls, and attempts to design systems that optimize safety and minimize the risk of error in complex environments" (AHRQ 2019). The International Ergonomics Association (IEA) develops a similar description of the concept: "Ergonomics (or human factors) is the scientific discipline concerned with the understanding of interactions among humans and other elements of a system, and the profession that applies theory, principles, data, and methods to design in order to optimize human well-being and overall system performance" (n.p., ratified by the IEA Council in 2000).

While human factors and interdisciplinary research target the same goals, and at face value appear analogous in general terms, there are important distinctions. Notably, human factors is rooted in psychological (cognitive ergonomics) and physiological (physiological ergonomics) perspectives, and, while a useful tool in the interdisciplinary toolkit, would alone afford a somewhat narrow view of the engineered system. The interdisciplinary perspective marshals a broader range of micro (e.g., psychology), meso (social psychology), and macro-level (sociology, political science, economics) perspectives and methodologies in an effort to more comprehensively understand and integrate the full range of factors governing the engineered system of interest. For example, in a hypothetical radar project, human factors may play a significant role in researching the colors and types of radar displays best suited for forecasting practice. However, broader social and cultural factors, such as how the public reacts to tornado warnings, how different sociodemographic groups respond to warnings, how local politics affects system deployment, and the potential economic costs of sheltering are all questions well beyond the scope an exclusive human factors approach.

Organization and Scope of the Volume

In Chaps. 1–3, we establish the context for discussing interdisciplinary warnings by exploring a comprehensive history of warning systems in the United States, identifying key milestones in warning technology, and we highlight key findings and debates in the social science literature on disaster warnings. This book offers a social science perspective on warnings contrasted against perspectives from other scientific disciplines and fields of practice. Summarizing the work of Milet and Sorensen (1990), as well as Lindell and Perry (1992), we present the current and prevailing perspectives among disaster scholars: far from simple alerts, warnings involve complex behaviors and multidimensional institutional coordination across numerous stakeholders—scientists, engineers, Emergency Operation Centers (EOCs), and, most vitally, the public. Findings will be discussed in terms of theoretical and methodological relevance for academics with great care taken to translate ideas, themes, and facts into practical terms for emergency managers, engineers, and meteorologists broadly tasked with developing strategies for the improvement of natural hazards warnings.

Chapter 4 builds upon the ideas presented in the first three chapters to identify and analyze common obstacles to successful warnings—namely, technological, policy, and psychological, social, and cultural obstacles. For instance, what scientific and technical limitations persistently limit detection and prediction of natural hazards? What roles do policymakers and politics—at the local, state, and federal levels—play in shaping warning policy? How do human or societal factors, such as disaster experience, socioeconomic status, and cultural beliefs further influence the warning response process? By cataloguing systematically potential barriers to successful warning, a background is set for later discussions of interdisciplinary warnings.

Chapter 5 introduces in further detail to the concepts of interdisciplinarity and interdisciplinary warnings. Key sections will present and discuss the history of interdisciplinary research in general, summarizing significant questions related to perspectives on end-user collaboration. The section also addresses major critiques of end-user design, weighing the strengths of the interdisciplinary approach against challenges not uncommon in collaborative research settings.

Chapter 6 provides an overview of the common problems encountered in interdisciplinary research, offering a short blueprint for the design of interdisciplinary research programs accordingly.

Chapter 7 summarizes and discusses present-day knowledge on the relevant groups, stake-holders, and end-users involved in the warning process—from top-level users, such as scientists to the general public. We present an extended case study based on the authors' work on ERC-CASA (Engineering Research Center for the Collaborative Adaptive Sensing of the Atmosphere). CASA's goal was to develop a new radar technology capable of more accurate and reliable predictions of severe weather conditions through an integrated, end-user approach in which relevant stakeholders—from radar engineers to emergency managers down to the general public—offer direction in system design. The project is an instructive, illustrative example given its character as an interdisciplinary effort involving cooperation between the sciences (physics, mathematics), engineering (computer science, radar engineering), social science (sociology, human factors), and practitioners (industry, emergency managers). The focus of this chapter will fall primarily on three end-user communities: the public, emergency managers, and scientists/ engineers.

Public: An overview of public response based on surveys from the 2006 to 2008 CASA study is presented. This research provides one of the few large scale interdisciplinary case studies of tornado warning response in the United States, applying contemporary social science models of the warning process to the examination of warning reception, understanding, and response.

Emergency Management: A summary and analysis of institutions' roles in the warning process is reviewed. Numerous experiments and case studies were conducted as part of the CASA project, ranging from qualitative interviews with emergency managers to natural experiments and surveys. Taken together, these studies portray emergency managers as a distinctly heterogonous community whose members differ considerably in terms of training, resources, knowledge, and capabilities. At the time, this research helped develop an end-to-end model for optimizing the CASA system for users with different backgrounds and skillsets. For instance, research determined that emergency managers encounter difficulty in making use of multiple information sources when making decisions, resulting in adjustments to system design.

Science/Engineering: This final section articulates the role played by natural and social scientists in an interdisciplinary project with a special focus on integrating meteorological approaches with social scientific theory in end-user design and practice.

Chapter 8 discusses trends that may hold consequences for the use and effectiveness of future warnings: notably, the rise of social media as a communication tool. This chapter discusses these emerging issues and reflects on consequences for an interdisciplinary perspective on warning systems.

In Chap. 9, the book concludes by bringing together the preceding information into a "lessons learned" and brief policy discussion. Interdisciplinary methods and theories of investigation in the field of disaster research, while growing in prevalence, suffer from a general lack of integration and lack an overall model for the development of scientific protocols. Insights from both the broader literature as well as the CASA project provides guidance for scholars planning interdisciplinary efforts or at present participating in multidisciplinary teams of researchers. The chapter concludes with a discussion of policy implications, as well as how the findings from this book can contribute to developing more effective warning systems that will have a significant and positive impact on disaster preparation and response, especially as it relates to severe weather, such as tornadoes, with the intended outcome of reducing the number of injuries and fatalities as a consequence of these devastating events.

Bibliography

Agency for Healthcare Research and Quality (AHRQ). (2019). "Human Factors Engineering." https://psnet.ahrq.gov/primer/human-factors-engineering. Accessed on 6/20/2022.

Ashley, W. S., Krmenec, A. J., & Schwantes, R. (2008). Vulnerability due to Nocturnal Tornadoes, Weather and Forecasting, 23(5), 795-807.

Bader, P., Boisclair D., and Ferrence, R. (2011). Effects of Tobacco Taxation and Pricing on Smoking Behavior in High-Risk Populations: A Knowledge Synthesis. *International Journal of Environmental Research and Public Health*, 8(11), 4118–39. doi: https://doi.org/10.3390/ijerph8114118.

Bayer, R. (2008). Stigma and the Ethics of Public Health: Not Can We but Should We. *Social Science & Medicine,* 67(3), 463–72. doi: https://doi.org/10.1016/j.socscimed.2008.03.017.

Brooks, H. (2009). "US Annual Tornado Death Tolls, 1875-Present – NSSL News." https://inside.nssl.noaa.gov/nsslnews/2009/03/us-annual-tornado-death-tolls-1875-present/.

Carson, N.J., Rodriguez, D., and Audrain-McGovern, J. (2005). "Investigation of Mechanisms Linking Media Exposure to Smoking in High School Students." *Preventive Medicine*, 41(2), 511–20. doi: https://doi.org/10.1016/j.ypmed.2005.01.002.

Chapple, A., Ziebland, S., and McPherson, A. (2004). Stigma, Shame, and Blame Experienced by Patients with Lung Cancer: Qualitative Study. *BMJ*, 328(7454), 1470. doi: https://doi.org/10.1136/bmj.38111.639734.7C.

Coleman, T., A. Knupp, Spann, K.R., Elliott, J.B., and Peters, B.E. (2011). The History (and Future) of Tornado Warning Dissemination in the United States. *Bulletin of the American Meteorological Society*, 92(5), 567–82. DOI: https://doi.org/10.1175/b2010BAMS3062.1.

Currier, J.. (2011). "Joplin Residents Ignored Tornado Warnings, Report Says." *St. Louis Post-Dispatch.* https://www.stltoday.com/news/local/metro/joplin-residents-ignored-tornado-warnings-report-says/article_66662dc8-e3b3-11e0-9262-0019bb30f31a.html). Accessed August 1, 2020.

Donner, W. (2007). An Integrated Model of Risk Perception and Protective Action. Doctoral Dissertation, Disaster Research Center (DRC), University of Delaware, Newark, DE.

Folger, P. (2011). "Severe Thunderstorms and Tornadoes in the United States." *CRS Report to Congress, Washington, D.C.*

International Ergonomics Association (IEA). What is Ergonomics? https://iea.cc/what-is-ergonomics/. Accessed 8/27/21.

Kar, B. and Cochran, D.M (eds.) (2019). *Risk Communication and Community Resilience*. 1st ed. NY: Routledge.

Kessler, E. (1970). Tornadoes. Bulletin of the American Meteorological Society, 51(10), 926–936.

Lindell, M.K. and R.W. Perry. 1992. Behavioral Foundations of Community Emergency Planning. Bristol, PA: Hemisphere Publishing Corporation

Lindell, M.K., Huang, SK., Wei, HL. et al. Perceptions and expected immediate reactions to tornado warning polygons. Nat Hazards 80, 683–707 (2016). https://doi.org/10.1007/s11069-015-1990-5.

Mileti, D.S. (2003). *Disasters by Design. A Reassessment of Natural Hazards in the United States*. Washington, D.C.

Mileti, D. S., and Sorensen, J. (1990). *Communication of Emergency Public Warnings: A Social Science Perspective and State-of-the- Art Assessment*. ORNL-6609. Oak Ridge National Lab., TN (USA).

Monárrez-Espino, J., Liu, B., Greiner, F., Bremberg, S., and Galanti R. (2014). Review of the Effect of Pictorial Warnings on Cigarette Packages in Smoking Behavior. *American Journal of Public* Health, 104(10), e11–e30.

Murphy, K. (2011). Many Failed to Heed Joplin Tornado Warnings, Report Says. *Reuters*, September 20.

National Weather Service (NWS). n.d. History of the National Weather Service. https://www.weather.gov/timeline. Accessed August 27, 2021.

NOAA. (2013). "US Annual Tornado Death Tolls, 1875-present." https://inside.nssl.noaa.gov/nsslnews/2009/03/us-annual-tornado-death-tolls-1875-present/. Accessed August 1, 2020.

Paul, B.K. and Stimers, M. (2014). Spatial Analyses of the 2011 Joplin Tornado Mortality: Deaths by Interpolated Damage Zones and Location of Victims, 6(2), 161–174.

Peacock, W.G., Morrow, B.H. and Gladwin, H. (1997). *Hurricane Andrew: Ethnicity, Gender, and the Sociology of Disasters*. London: Routledge.

Quarantelli, E. L. (1954). The Nature and Conditions of Panic. *American Journal of Sociology*, 60 (3), 267–75. doi: https://doi.org/10.1086/221536.

Stokoe, R.M. (2016). Putting People at the Centre of Tornado Warnings: How Perception Analysis Can Cut Fatalities. *International Journal of Disaster Risk Reduction*, 17, 137–53. doi: https://doi.org/10.1016/j.ijdrr.2016.04.004.

Tversky, A. and Kahneman, D. (1974). Judgment under Uncertainty: Heuristics and Biases. *Science,* 185(4157), 1124–1131.

US Department of Commerce, National Oceanic and Atmospheric Administration. "NOAA 200th Feature Story: History of Tornado Forecasting." (https://celebrating200years.noaa.gov/magazine/tornado_forecasting/).

Chapter 2
Historical, Technical, and Social Evolution of U.S. Warning Systems

Before discussing modern warning systems, it is first necessary to understand the history of meteorology generally and severe weather research specifically. It must be born in mind that scientific fields rarely evolve in isolation, but rather emerge and develop alongside the social and cultural institutions of an era, with the atmospheric sciences standing as no exception to this principle. Indeed, the growth of meteorological science, necessary for the development of warning systems, is itself a reflection of our cultural diversity, evolving social structures, growing urbanization patterns, and the drive towards increasing technological advancements. In a less complicated past, warning systems were once as modest as a town crier, and today are as sophisticated as space-based sensing driven by artificial intelligence. The progress in the development of these systems reflects the successes of past generations and point to the challenges and potential of the future.

This chapter provides a critical foundation for later discussions by providing an historical and social overview of meteorological science and warning systems. First, we trace the history of warnings through the early days of meteorology, in which observations were often made without the aid of instruments, through the present period, in which scientists are close to realizing near real-time global monitoring of the natural world. Following this discussion, we elaborate on the range of modern technologies and policies that encompass the current warning system.

The Early History of Severe Weather Warnings

Weather was and remains a source of great fascination and awe among scientists, practitioners, and laypeople alike. Although attempts to understand and predict the weather date back to ancient times (Hellmann 1908), the science of meteorological observation and prediction, and in particular severe weather forecasting, saw few meaningful advances prior to the early-to-mid-twentieth century. Early efforts to develop meteorology as a scientific discipline, such as was seen with Finley's

W. Donner et al., *Severe Weather Warnings: An Interdisciplinary Approach*,
https://doi.org/10.1007/978-3-031-05031-2_2

influential work on tornado predictions in the 1870s, did much to advance our knowledge of weather conditions, but failed to earn the kind of widespread institutional acknowledgement and support out of which to establish meteorology as a formal discipline (Fleming 1990). With rare exception, the investigation of severe weather stagnated until the 1940s, a trend in large part due to the transfer of meteorological operations from scientific institutions (such as the Smithsonian) to the U.S. Signal Corps, which held little interest in the basic science of meteorology. Financial and political pressures moved weather research towards more pragmatic goals for most of history. From the mid-twentieth century onward, however, a renaissance in the field of meteorology brought forth progress in the realm of observational, theoretical, and applied meteorology and, consequently, warning systems. Thus, we divide the early history of warnings into two distinct periods. The first, in which warnings were treated with suspicion and the second in which technology began to rapidly advance, thus opening a path for later interdisciplinary research efforts.

Meteorology of the Classical and Renaissance Eras

The weather has long since captured the attention of societies dating to the early classical period. The Greeks established much of the conceptual foundations for understanding the physical world. Perhaps the earliest known work in meteorology, Aristotle published his four-book "Meteorologica" in 350 B.C., followed by Hippocrates "Air, Waters, and Places" in 330 B.C. But the natural world provided the earliest warning systems. The New Testament references the use of a red sky in the morning as a common predictor of storms (Matthew 16: 2–3). Some indigenous peoples still follow the behavior of animals to warn of tsunami and earthquakes, and modern science confirms the validity and value of these natural warnings. But human ingenuity would be key to further warning system development. One clever early example, which serves as an illustration of perhaps the first warning system, was a seismoscope invented in 132 AD by Chinese astronomer Zhang Heng during the Han Dynasty (25–220 AD) (*Ancient Origins*). This came in the form of a large bronze vase 6′ in diameter with eight dragons facing outward in each of the principal directions. Each dragon held a small bronze ball in its mouth. Should an earthquake strike with the most delicate of tremors, the ball drops from the dragon's mouth into the mouth of a waiting toad below, the sound of which provided its owners with the slightest of advanced warnings. The ball that falls points to the direction of the earthquake's epicenter. Modern reconstructions of the vase have proven the validity of the design, a warning system clearly years ahead of its time.

But it was not until the fifteenth century that instruments were built to directly observe and document the natural world in a more methodical way. Cardinal Nicholas of Cusa (1401–1464) invented the hygrometer in 1450, Galileo Galilei (1564–1642) developed the first thermometer in 1593, and Evangelista Torricelli (1608–1647) designed the first barometer in 1644. Korean King Sejong

(1397–1450) developed the first rain gauge ("Cheugugi") in 1442 (*London Korean Links*). By monitoring data from these sensors, trends could be used to predict the future atmospheric state. These and additional in situ weather sensors continue to be developed and refined.

1870–1948: Early Science, Institutional Support, and Stalled Progress

The end of the classical and medieval renaissance eras met with growing interest in the meteorological sciences. Nevertheless, would-be forecasters faced significant obstacles towards the development of meteorology as a formal scientific field. Most notably in this regard, the nineteenth century witnessed few scientific patronage organizations, the absence of which crippled efforts at developing meteorology into an institutionalized field of science (Fleming 1990). There was simply little support for science in general and meteorology in particular. As a result, during this period most research was conducted by the well-meaning but poorly-resourced amateur scientist in many colonial contexts (Williamson 2015).[1] Perhaps most emblematic of this trend, Benjamin Franklin's work, though limited by the primitive technologies of the time, successfully surveyed weather phenomena and recorded the movement of storms (Abbe 1906). Although of little practical use by today's standards, such results had the important effect of fostering interest in the science of weather observation. Still, however rigorous and systematic their effort, without the aid of modern equipment, Franklin and his contemporaries could do scarcely more than document basic meteorological phenomena, which ultimately held limited practical forecasting value. The development of meteorological science, much less warning systems, would remain limited in scope and practical utility under such conditions. However, a boon to science in the U.S. was on the horizon.

This boon took the form of unprecedented growth in U.S. domestic patronage institutions, the previous absence of which stymied scientific progress in meteorology and other fields. Organizations such as the Smithsonian Institute were the first to offer standard support in the form of finances, resources, and technology for scientific inquiry, the impacts of which cannot be overstated. Historians note that the Smithsonian Institute, along with the Joint Committee on Meteorology and Franklin Institute, "had significant impact on meteorological theory, providing mechanisms for professional entry and advances in science and better control over

[1] In comparison, patronage was a well-established practice in both the arts and science in Europe, offering a means to build social status among both patrons and scientists alike. For instance, the House of Medici funded Galileo not so much in the spirit of scientific discovery, but in order to build prestige, and, in so doing, further consolidate power (Westfall 1985). Notably, the moons of Jupiter originally bore the Medici namesake. Intent aside, such funding nevertheless advanced scientific thought, leading, for instance, to the invention of the telescope. No similar system existed in the United States.

the quality and amount of data" (Fleming 1990: 173). A move towards organized investment in the sciences would result in the establishment of a network of amateur meteorologists operating under the leadership of Joseph Henry, the first Secretary of the Smithsonian Institute, who would accordingly go on to coordinate and communicate spotter observations to newspapers. This was possible thanks to the widespread acceptance and use of Morse's telegraph technology and the concomitant and very rapid creation of the infrastructure for its large-scale deployment. This revolution in communication made the practical application of forecasting science a reality. Thus, the growth of patronage, combined with the development of communications infrastructure in the mid-nineteenth Century, resulted in advances in weather observations and theory as well as their application in forecasting practice. The stage was set for a greater focus on severe weather, and, by extension, the development of warning systems.

Sadly, however, this was all to be short lived. Changing political alignments and historical events would serve to draw attention away from the study of the weather, stalling the development of meteorological science. Interest in the weather for the sake of scientific discovery declined in the tumult that characterized the U.S. mid-1800s, which faced "a series of calamities—war, fire, financial panic, and changing cooperative arrangements— [that ultimately] dealt severe blows to the Smithsonian meteorological project" (Fleming 1990: 162). In the face of waning support, meteorology slowly abandoned its more academic focus as weather operations were placed under the U.S. military Signal Corps. The Signal Corps gradually annexed and accordingly redirected U.S. meteorological operations towards more applied forecasting activities (Spilhaus 1950).

Creation of the Weather Bureau

In 1871, for the U.S. weather community, a series of policy and organizational changes aimed at encouraging applied forecasting at the federal level led to a rapid decline in an autonomy that had previously enabled the pursuit of basic science within the U.S. meteorological community. No federal weather institutions existed at the time. This would change, however, when President Grant placed a newly-created Weather Bureau under the U.S. Army's Signal Corps with the intention "to provide for taking meteorological observations at the military stations in the interior of the continent, and at other points in the States and Territories. . .and for giving notice on the northern lakes and on the seacoast, by magnetic telegraph and marine signals, of the approach and force of storms" (NOAA n.d.). The message was quite clear: meteorology would, with exceptions, primarily exist to serve the interests of the military and would, as a result, focus less on basic weather and climate theory. The resolution to place the Weather Bureau under military governance would have significant consequences for the development of U.S. meteorology. Under the authority of the Army Signal Corps, meteorologists saw basic research sacrificed in favor of operational forecasting aimed towards supporting commerce, agriculture, and, as one might infer, military operations in an era later heralded as the "dark ages" of weather research (Grice et al. 1999: 1345). Notably, despite a growing focus on practical forecasting, little headway was made in the area of severe weather meteorology, since the prevailing sentiment at the time was that "tornado forecasts would do more harm than the tornado itself." (Grice et al. 1999: 1345; see also Bradford 1999). As a result of declining scientific focus, as well as entrenched skepticism regarding the reliability of tornado forecasts, the development in the theory and practice of warning systems lay dormant for decades. Nevertheless, great strides were made in the practical elements of operations, and technological advances led to greater confidence in the potential for forecasting weather. Perhaps most importantly, with the invention of the telegraph, there was a growing sense that lives could be saved with the aid of mass communication, and so government policy took full advantage by formally adopting its use.

Changes would come about after roughly two decades. As the nineteenth Century drew to a close, the Harrison administration transferred the Weather Bureau from the military-run Signal Corps to the civilian-run Department of Agriculture in 1890. Through the following years, the Weather Bureau would further expand its mission as capabilities broadened, issuing flood warnings in 1891, and establishing a hurricane warning network across the West Indies in 1898. Following the sinking of the Titanic in 1912, international ice patrols began, and the first fire weather forecasts were issued that same year (NOAA n.d.). Despite its expanding role in predicting hurricanes, floods, and fires, however, the Weather Bureau stubbornly refused to deal with tornadoes in particular. In 1886, the Weather Bureau went as far as to ban the word "tornado" from any of its alerts, fearful that the very mention of the term would incite public panic. It is also relevant to mention that under civilian management the Weather Bureau saw some freedom to pursue basic science, although the specific mission of the Department of Agriculture ensured that most meteorological efforts would remain application focused, supporting the farming industry, and, later, business and shipping under the Department of Commerce. It would be some time before political entities would allow for public issuance of tornado warnings.

1948: Lifting the Ban on Tornado Warnings

Despite advances in the science of severe weather, as well as some success in warning hurricanes and floods, officials remained hesitant to issue public tornado warnings. A "proof of concept" that warnings could at once be reasonably reliable and at the same time avoid triggering panic would be necessary to overturn the suspicion with which tornado warnings were regarded during the period. Such proof came in the form of a highly-visible and highly-successful warning at an Army base in Oklahoma immediately following the end of World War II. On March 25, 1948, Captain Robert C. Miller and Major Ernest J. Fawbush, two meteorologists stationed at Tinker Air Force Base in Oklahoma City, issued what would become the first successful tornado forecast in history (Grice et al. 1999). Recognizing the need for a system to protect military and civilian assets in a region plagued by severe weather, base commanders charged the two forecasters with developing a method for forecasting and warning tornadoes, a problem for which no reasonably reliable solution existed at the time. Through an examination of historical meteorological data, as well as an analysis of an earlier tornadic event just days earlier on March 20th, Miller and Fawbrush would later go on to issue the first successful tornado forecast. The tornado caused over $6 million in damage and several injuries, but the tornado warning led to safety measures ahead of the storm which minimized losses. This accomplishment is all the more extraordinary due to the relative sparsity of weather data and information at the time, thereby making Fawbrush and Miller's achievement all the more striking. They enjoyed the use of neither a highly sophisticated Doppler radar nor satellite technology of the modern era, nor were the modern prediction algorithms available to make sense of the limited data to which they had access. A precedent, nevertheless, was now set: tornado warnings were now a realistic tool to save lives.

The Tinker Warnings are of great significance because while the late nineteenth century and early twentieth century period saw progress on tornado prediction through Finley's Smithsonian-backed research, progress was limited due to the tornado warning ban (Bradford 1999). The ban on tornado warnings remained firmly entrenched operationally for some time (https://celebrating200years.noaa.gov/magazine/tornado_forecasting/#early). But following the success of Fawbrush and Miller's forecast, policymakers began to rethink this prohibition. It was around this time, from about 1942, that a second critical element of warning systems began to develop: spotter networks (Coleman et al. 2011). While general storm spotting networks existed since 1890 (https://www.weather.gov/coop/Overview), by 1950, Doswell (1999: 546) reports that tornado spotter networks were well on their way to being "commonplace," adding that "this evolution followed rather directly in the wake of the inception of public tornado forecasts, permitting the alerting and deployment of spotters in advance of threatening weather situations."

Along with the success at Tinker and growing spotter networks, three significant tornadic events in 1953 increased public pressure for an improved national tornado warning program. Tornadoes in Waco, Texas (119 deaths), Flint, Michigan (115 deaths), and Worcester, Massachusetts (94 deaths) caused devastating damage within 1 month's time. Fortuitously, in 1952 the Weather Bureau established a dedicated severe weather unit, known as the Severe Local Storms Unit (SELS) (http://www.spc.noaa.gov/history/early.html), a precursor to the Storm Prediction Center (SPC) (http://www.spc.noaa.gov/faq/). With the warning ban formally lifted in 1950 (Coleman et al. 2011), SELS enjoyed great freedom to issue tornado forecasts that closely mirrors today's tornado watches in scope and aim. That same year, Weather Bureau Director Francis Reichelderfer (Bradford 1999: 78) stated: "There is no regulation or order against the forecasting of tornadoes. Whenever the forecaster has a sound basis for predicting tornadoes, the forecast should include the prediction in as definite terms as the circumstances justify." In so rejecting the previous injunction against tornado warnings, Reichelderfer lifted the ban that was in place for nearly half a century, opening the door not only for operational warnings, but also for institutionalized warning research, thereby ushering in a renewed era in our understanding of severe weather.

Today, a majority of U.S. severe weather research is guided by formalized institutions such as government laboratories and universities. The National Science Foundation (NSF), National Oceanic and Atmospheric Administration (NOAA), and the National Aeronautics and Space Administration (NASA) provide the bulk of the nation's funding and guidance for atmospheric research. The National Severe Storms Laboratory (NSSL), the National Center for Atmospheric Research (NCAR), the National Hurricane Center (NHC), and the Atlantic Oceanographic and Meteorological Laboratory (AMOL) continue to be the national leading centers for atmospheric research. Atmospheric research is now a central tenant of the nation's scientific enterprise.

Modern Warning Systems: Evolution of Core Technologies and Practices

Detection and Prediction of Severe Weather

Starting from the 1950s, cutting-edge computer and engineering technologies combined with basic theoretical science principles and experience created the largely successful, technically-driven warning systems in place today. Many of the advancements are rooted in past innovation and discovery. Indeed, advances in warning systems reflect each era's understanding of the physical world, technology, and engineering abilities. Earlier periods saw the introduction of the seismograph. Networks of these sensors were soon deployed, which led to a greater understanding of the physical world. Concepts such as the hydrological cycle, high and low pressure systems, and the translation of weather systems began to be better understood. Greater exploration of the world led to more observations of tornadoes, hurricanes, volcanoes and tsunami. These observations led to improved sensors, which led to new concepts; these new concepts led to new and improved sensors; and the cycle continues through the present. Warning technology has advanced from individual sensors, networks of sensors and peoples, and then through remotely-sensed observations, telecommunications, and now computer and satellite systems. This has led to the modern, integrated warning system.

A modern, integrated warning system is comprised of four basic elements: (1) forecast; (2) detection; (3) dissemination; and (4) response (Doswell et al. 1999; Leik et al. 1981). This section explores the historical evolution of these technologies that undergird the integrated warning system. Chronologically, detection sensors were deployed and studied long before forecasts could be generated, and it wouldn't be until the second half of the twentieth century that numerical weather prediction (NWP) models were operationalized. As such, in the following sections.

we explore the technological development of detection systems first, followed by forecast technologies, and then will end with a review of current dissemination technologies and public response options.

Sensor Networks

As important as individual sensors are for understanding trends, it wasn't until networks of sensors began to be deployed simultaneously over broad geographic areas that we were able to sample observations in time and space. Mapping data from these sensors spurred tremendous progress in understanding the evolution of weather systems, which translated to improved applications and eventually warnings. During the 1880s, English geologist John Milne (1850–1913), together with colleagues James Ewing (1855–1935) and Thomas Gray (1850–1908), developed and popularized the use of modern seismographs in Japan (MIT n.d.) In 1895, John Milne returned to England where he proposed and constructed a global network of seismograph stations (Agnew 2002). During the eighteenth Century, early efforts at deploying networks of weather stations were attempted by the British Royal Society, the Royale Societe de Medicine de Paris, and the Academia de Cimento (Glaser and Riemann 2009). However, the Meteorological Society of Mannheim was the most successful as it was the first to standardize stations; they set up an international network of 39 stations from North America to Russia, collecting observations between 1781 and 1792 (Aspaas and Hansen 2012). Beginning in 1814, the United States recorded weather data regularly (three times daily) using standardized equipment across its military posts (Landsberg 1964). These observations were ordered by the Surgeon General of the Army with the foremost interest in developing relationships between climatology and health. The success of this network spurred additional networks across the country.

In an effort to solve "the problem of American storms", the first Secretary of the Smithsonian Institute, Joseph Henry (1797–1878), proposed in 1847 an expanded nationwide weather network. The Smithsonian approved a 100 station network with the goal of placing one site per 10,000 square miles. Its first year saw 150 volunteers, and by 1860, the network had expanded to 500 stations across the country (NOAA n. d.). Daily observations were mailed to the Smithsonian monthly. In 1890, observing responsibilities were passed to the new U.S. Weather Bureau within the Dept. of Agriculture. In the same year, the Cooperative Observer Program (COOP) was started, which uses volunteers to collect daily weather observations. The COOP has been wildly successful; it remains active today, and is now the largest weather network in the country with over 8700 stations in operation (US Department of Commerce, NOAA n.d.-b).

Despite the scientific value offered by many of these innovative networks, most were still of limited value for most applications. Systems were deployed for research purposes only, or for developing long-term climatologies for agriculture and health. The inherent limitation for these early networks was slow communication. Prior to the nineteenth century, observations had to be written down by hand and then collected through the mail. The transformational breakthrough in communications was the invention of the telegraph in 1833 by Samuel Morse (1791–1872), and its impact on meteorology, and for warning systems in particular, cannot be overstated.

The telegraph's rapid communication allowed, for the first time, real-time applications—warnings and even forecasts downstream. As part of the Smithsonian's weather network, Joseph Henry contracted for 20 telegraph stations across the country to report daily observations of sky conditions, precipitation, and wind direction. These were compiled by Henry into a national daily weather analysis map, which he then posted in a prominent area of the Smithsonian museum (Smithsonian Institute 2017). Admiral Robert FitzRoy (1805–1865), perhaps best known as the captain of the HMS Beagle, was among the first to begin using transmitted weather data for warnings. Troubled by the number of fatal shipwrecks along Britain's coasts, in 1860, FitzRoy began collecting real-time weather observations, and then once analyzed, telegraphed storm warnings to various harbors. He began issuing general forecasts a year later. While heavily criticized for the inaccuracy of his forecasts, FitzRoy's team would persist, later becoming the British Met Office (Moore 2015).

The turn of the Twentieth Century brought with it a great expansion in sensing capabilities and engineering. In 1909, the Weather Bureau began to use weather balloons for measuring upper-air information, and by 1931 the use of aircraft observations had become routine. The U.S. Navy developed automated weather stations in 1939, but it would not be until 1989 that the National Weather Service (NWS) launched the Automated Surface Observing System (ASOS) to replace manual recordkeeping routinely. Over 900 ASOS stations are now deployed across the U.S. Mid-century regional networks of stations (referred to as mesonets) were first deployed to monitor more localized, mesoscale features, such as the 13-station mesonet operated by the FAA between 1965 and 1967 in and around Atlantic City, NJ (Entrekin et al. 1969). As technology progressed, states began to deploy their own mesonets, and now nearly half of all states have some form of a mesonet with over 1600 mesonet stations operating as of 2017 (Mahmood et al. 2017). Ground, air, and hydrologic in situ measurements continue to expand across the globe.

Development of Remote Sensing and Radar Technologies

In situ sensors are those sensors immersed in what they are measuring. However, most natural hazards cannot be directly measured with in situ instruments, but instead must be inferred from observations collected in the vicinity. Indeed, there are simply not enough in situ sensors to adequately monitor all hazards all the time. For example, a developing tornado cannot be directly measured with an anemometer unless one is extremely 'lucky'; instead, such a hazard usually must be inferred from a network of surface and upper air data surrounding the storm, and then only if you have a sufficiently dense network. The development of 'remote sensing' systems greatly enhanced our ability to monitor and predict the natural world.

First known as Radio Detection and Ranging (RaDAR), radar was a British innovation that grew out of radio physics. An electromagnetic wave is emitted in a particular direction, the signal bounces off of a fixed object, and the time it takes for the signal to return to the radar is used to determine the distance to that object. Sir Robert Watson-Watt (1892–1973) invented radar in 1935, and by 1939 Britain had deployed a full network of radars, known as the British Chain Home (CH) network, to monitor German aircraft attacks. Early users quickly realized that precipitation interfered with the signal. By the end of the war, the U.S. military had small radar networks deployed for weather operations in Panama and India. This new ability to monitor 'sferics' using remote sensing provided an invaluable new tool for meteorologists to observe clouds, precipitation. and wind above ground in three-dimensions.

Following WWII, weather radar was soon adopted for civilian use, and their value for monitoring natural hazards was quickly realized. Several hurricanes were observed by radar in 1946, and the first tornado was observed by radar in 1953. The use of weather radar was seen as a critical element in early warning systems. In 1955, a consortium of university, local, state and federal partners formed the Texas Tornado Warning Network, a system of 17 weather radars deployed for early warning detection. In response to several highly publicized deadly tornadoes in 1953 and a series of impactful hurricanes in 1954 and 1955, the federal government commissioned its own network of 31 Weather Surveillance radars in 1957 (WSR-57). These were largely deployed along the nation's coasts and across tornado alley to monitor the hurricane and tornado threats, respectively. The network was expanded and later upgraded in the 1970s with the WSR-74. In the meantime, military radars began utilizing Doppler radars to determine the speed at which objects (planes or storms) were moving. In 1988, a joint partnership between the NWS, DoD and FAA commissioned a next generation radar (NEXRAD) network of 165 Weather Surveillance Radar—1988 Doppler (WSR-88D). These were deployed across the country. A second network of 45 Terminal Doppler Weather Radar (TDWRs) were installed at select airports and utilized a shorter wavelength to detect wind shear for aircraft. The Doppler capability allowed wind speed to be directly measured. The WSR-88D network is still in use, recently upgraded with dual-polarization capability, which allows the size, type and behavior of aerosols to be observed. This capability has improved the understanding and warning of flash floods, severe winds, hail and tornadoes.

New radar technologies continue to develop rapidly. Currently, operational radars still must scan mechanically; however, the military has developed phased array radars (PARs) where radars scan digitally. This allows for much faster scanning rates, potentially providing faster warnings for severe thunderstorms and tornadoes. A second advancement is the operationalization of networks of small (X-band) radars, such as CASA (e.g., http://www.casa.umass.edu/). The overlapping coverage from multiple radars enables higher resolution monitoring of wind and precipitation in three dimensions. Integrated radars can scan collaboratively, working together to optimize the observational space. These radar networks are particularly useful in urban areas or complex terrain, where additional radar coverage is needed at low levels.

An arguably even greater advancement to come out of the twentieth century was the placement of remote sensing platforms in space. Satellites revolutionized how we see weather. Instead of viewing weather only from the ground up, we're now able to monitor the weather from above, like watching ants at a picnic. The world's first weather satellites were launched by the U.S. in 1960. Known as TIROS I AND TIROS II, they collected real-time imagery of cloud cover and solar radiation, providing critical information for forecasts and warnings. TIROS III followed in 1963. Space exploration continued with the Apollo missions, and in 1975, the first Geostationary Operational Environmental Satellite (GOES) satellite was put into orbit. The GOES series of satellites provide the operational weather community with continuous weather coverage from space, with GOES-16 and GOES-17 now in orbit. Numerous other satellites provide high-resolution imagery, vertical soundings, and data on clouds, ice, oceans, wildfires, ozone, soil moisture and temperature, vegetation, and land use data among others.

Within 200 years, the world has progressed from recording weather data by hand and mailing it in to global networks of automated surface, hydrologic, and space-based systems instantly transmitting data to central hubs for instant processing, where they are often then assimilated into high-resolution analyses and numerical weather prediction (NWP) models for ready use by emergency management worldwide.

Forecast Technologies: Numerical Weather Prediction

For some warning scenarios, a relatively long lead time is needed in order to provide adequate time to shelter. For example, ships at sea often need hours if not days advance notice in order to steer clear of large tropical systems, and large cities need days of advance notice to evacuate ahead of advancing hurricanes. To provide such long lead times requires detailed, accurate and reliable NWP.

Numerical weather prediction is comprised of several steps. First, weather observations are collected in space and time from a variety of weather sensors, as discussed previously. Next, an analysis is created as observations are analyzed to a grid at a given point in time. Finally, a set of physical equations are applied to the grid that "move" or predict the translation of those observations in space and time. The more accurate the initial data and the physical equations, the more representative and accurate the forecast. Initially, numerical models were run at very coarse grids due to limitations in computational power, but NWP resolutions have improved dramatically as computers and networking have progressed.

While meteorologist Vilhelm Bjerknes (1862–1951) was the first to propose the idea of numerical weather prediction, British mathematician Lewis Fry Richardson (1881–1953) is now considered the "Father" of NWP (NASA 2002). In an effort to apply a finite differencing technique he had developed, Dr. Richardson gathered weather data from across central Europe for a single day (May 20, 1910), and then spent several months crunching the numbers using several atmospheric equations to produce a single 6-h forecast. Although the resulting forecast was physically impossible, he published the results in a 1922 book titled, "Weather Prediction by Numerical Process". Despite his failure, he showed how it could be done and what was required. Since this was in the days long before computers, Richardson suggested that NWP could be done in real-time if you could utilize 64,000 people, all crunching numbers in a room together simultaneously. Fortunately, the advent of the computer negated the need for hiring quite so many assistants.

Indeed, the creation of the computer was intricately tied to the development of numerical weather prediction. In 1948, the Princeton Institute for Advanced Studies, led by mathematician John von Neumann (1903–1957), began building a computer with the goal of testing it by running a primitive numerical weather prediction model (NWS n.d.; NASA 2002). That first computer was known as the Electronic Numerical Integrator And Computer (ENIAC). Meteorologist Jule Charney (1917–1981) led the creation of the numerical model, and by 1950 Charney's team had generated a successful 24-h forecast over North America. In 1954, the Joint Numerical Weather Prediction Unit was formed, which began producing twice-daily continental U.S. (CONUS) forecasts. For comparison with today, the models then had a grid spacing of several hundred kilometers but now has a grid spacing of 3 km in the highest-resolution model run operationally [the High-Resolution Rapid Refresh (HRRR)].

Successive operational models progressed rapidly as computer power increased exponentially; the NWS introduced the first regional model in 1971 and the first global model in 1974 (US Department of Commerce, NOAA n.d.-a). During the first few decades of NWP, models were largely run at synoptic to global scales due to the limitations in computing power; however, by century's end mesoscale models, such as the Advanced Regional Prediction System (ARPS), MM5, and Weather Research and Forecasting (WRF) model, would be run routinely with grid scales of only tens of kilometers and able to accurately forecast severe storms hours in advance. These models continue to be run at ever finer spatial resolutions at faster speeds and even within real-time warning situations. The next step is to integrate very short-term (1 h) NWP directly into the warning process in what is called "Warn-on-Forecast"; for example, tornado warning areas may soon be drawn based on real-time NWP predictions.

In an effort to better understand model predictability and variance, multitudes of numerical models are now run simultaneously to create an "ensemble". Because every model has certain biases and errors, applying post-processing to the output from a suite of weather prediction models can yield a more accurate forecast than from any single model member. Most importantly, the spread among model members provides a measure of forecast uncertainty, a key parameter for emergency managers wishing to understand the probabilities for high-impact events.

Within a mere 100 years, numerical weather prediction has progressed from crunching numbers by hand for months at a time for a wildly inaccurate 6-h forecast to solving trillions of calculations per second by computer for a more accurate, longer, higher-resolution prediction. Numerical model predictions are now a critical element to generating forecasts of air pollution, allergens, hurricane evolution, severe storm outbreaks, winter storms, space weather, tsunamis, and volcanic eruptions. With continued progress in computing power, telecommunications, and artificial intelligence, we can expect to see further improvements in model accuracy, resolution, and reliability.

Spotter Networks

While networks of people are not a technology per se, the harnessing of people networks for detection has proven to be a persistently effective warning tool. Prior to today's modern warning system, tornadoes were particularly deadly; per capita deaths from tornadoes during the mid- and late nineteenth century were 20 times the rate of today. To address the problem, the U.S. Army Signal Corps initiated in 1882 a weather program to better understand and warn on tornadoes. Led by Sgt. John P. Finley (1854–1943), he enlisted volunteer tornado "spotters" from across the central and eastern U.S. Based on the data collected by the network, by 1884 Finley had started issuing rather successful 'tornado alerts', a forerunner of today's Tornado Watch (Coleman and Pence 2009). The number of tornado spotters eventually grew to 2376 by 1887. However, following the government ban on the use of the word "tornado" in 1886, the spotter network appears to have fallen into disuse after its peak.

In 1938, the Weather Bureau re-approved the use of the word "tornado" for warnings, though tornado forecasts were not approved until over a decade later in 1950. By 1942, the Weather Bureau had again revitalized the use of storm spotter networks, in this case to monitor severe weather around military posts (Doswell et al. 1999). By 1945, nearly 200 spotter networks were active nationwide. However, it wasn't until tornado forecasts (Watches) began to be issued that spotter networks could be given a much clearer sense of the timing and location to be on the alert. With the creation of SELS, in 1952 the NWS began to issue official tornado "forecasts" to the general public. These forecasts designated large areas (up to 100,000 km^2 in size) where tornadoes could occur within the following 7 h (Coleman et al. 2011). The term "Tornado Watch" was standardized in 1965 to minimize confusion.

On 11 April 1965, 271 people died in what became known as the Palm Sunday tornado outbreak. While forecasts were relatively good, a post assessment team found that the public generally either did not receive the warning or did not know exactly how to respond to it (Doswell et al. 1999). In response, the term "Tornado Warning" was officially adopted, again to minimize public confusion. The federal government also developed the Natural Disaster Warning system (NADWARN). NADWARN was designed to better coordinate all relevant emergency management agencies, facilitate more efficient dissemination, and improve public response. One outcome of this proposal was the formation in 1969 of SKYWARN (https://www.

skywarn.org), a designated NWS program to train volunteer members of the general public to become storm spotters. A several hour training course is provided, which educates students on storm identification, communication, and safety. Recertification is encouraged every 2 years. The program has since trained between 350,000 and 400,000 participants (US Department of Commerce, NOAA n.d.-c).

Despite the many modern observing tools, spotter networks remain critical. Spotters provide ground truth, confirming what is often only suspected from radar or satellite. Spotters require regular training and can occasionally submit false reports. But reliable experienced spotters are essential for providing real-time reports from the field. Spotters often report directly to the NWS using either phone or ham radio. Today, many storm chasers double as storm spotters and emergency managers.

Warning Communication

In addition to revolutionary advances in the detection and prediction of severe weather, parallel improvements in the technologies and practices of warning communication have also taken place. Indeed, the greatest strength and greatest challenge of today's modern warning system is its diversity. We have learned over the years that "no size fits all", and that is as true of warning dissemination systems as it is with shirt sizes. And while we may be tempted to simplify our systems, whether in an effort to reduce costs or reduce management complexity, every warning component has its advantages and drawbacks and every option is seemingly irreplaceable. Legacy systems often have a uniqueness that cannot be found by newer means, and yet many modern systems provide new advantages that are equally important. This section explores many of the dissemination technologies in use today, explores their advantages and disadvantages, and examines where they might fit into the warning system of the future.

Storm-Based Warnings and Impact-Based Warnings (IBWs)

In recent decades, significant changes have been made to how warnings are created, communicated and displayed. This change in thinking was largely due to two factors. First, more sophisticated technology (e.g., radars), algorithms and models dramatically improved the reliability and accuracy of detecting, monitoring, and predicting severe weather at ever higher resolutions. Second, a growing body of research found that warning false alarms decreased trust (i.e., the "cry wolf effect") which led to higher casualty rates (Simmons and Sutter 2009) and that false alarms caused a significant waste of time and money (Sutter and Erickson 2010). In other words, minimizing deaths and injuries, as well as economic savings, could be realized by reducing the number of false alarms.

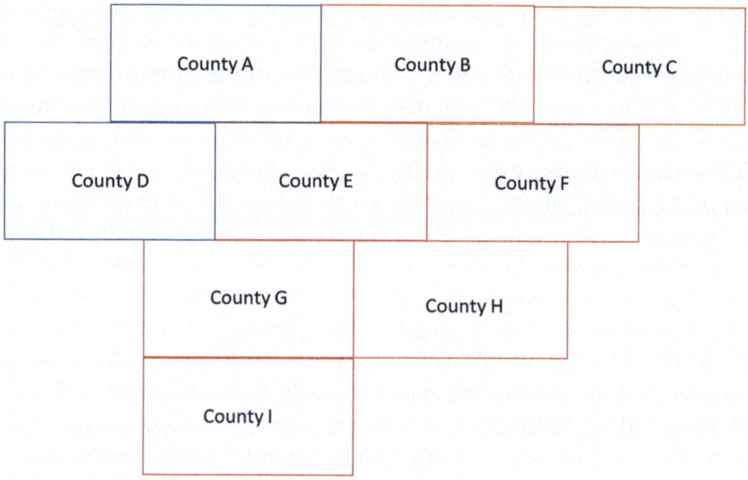

Fig. 2.1 County-based warnings. All counties, and all individuals within those counties, in red are considered warned and at-risk of an imminent severe weather threat

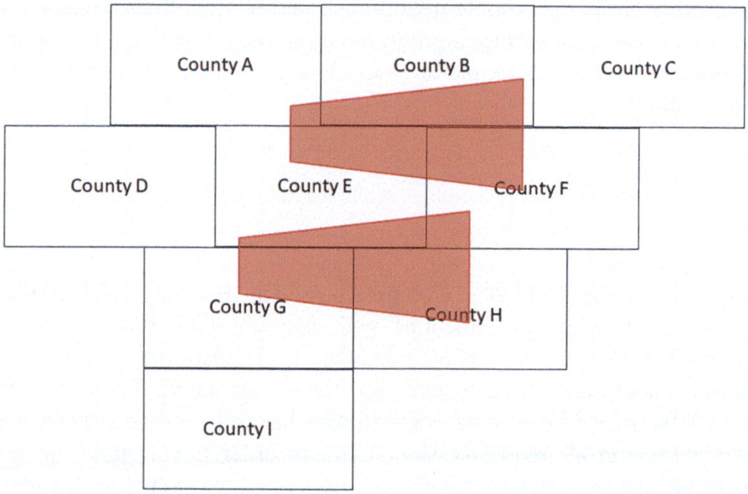

Fig. 2.2 Storm-based warnings. Instead of warning entire counties, only the portion of counties in the red cones are considered warned and at-risk of an imminent severe weather threat

To address these factors, in 2007 the NWS implemented major changes to how warnings are issued. Instead of warning entire counties (i.e., "county-based" warnings; see Fig. 2.1), warnings are now issued based on individual cell movements (i.e., "storm-based" warnings; see Fig. 2.2). The size and shape of the storm-based warnings are created as a function of the storm size and projected path. Overall, storm-based warning has allowed for a reduction in the size of the area warned while retaining high accuracy directly ahead of the storm path, thereby reducing the "false alarm rate" by reducing the footprint of the area warned. Coincident visualization shows a much-reduced warning area with the new storm-based warning paradigm. While more precise, and allowing for greater warning specificity, storm-based warnings present some particular challenges as well, as will be discussed later.

In addition to the dynamic changes to the physical dimensions of the warning, an evolution in the verbalization of written warnings would take place soon thereafter. In 2012, "impact-based warnings" (IBWs) were first tested and deployed. Instead of simply describing the size, location, and characteristic of a weather event, impact-based warnings sought a more evocative language in order to communicate risk with a greater sense of psychological gravity. For instance, an example of a tornado warning on the U.S. National Weather Service (NWS) reads: "You are in a life threatening situation. Flying debris may be deadly to those caught without shelter. Mobile homes will be destroyed. Considerable damage to homes. . .Businesses and vehicles is likely and complete destruction is possible" (https://www.weather.gov/impacts/examples).

In other words, while prior warnings might say something to the effect of "Large F2 tornado 5 miles from town X," IBWs attempt to evoke fear and concern with descriptors such as "deadly," "catastrophic," and "complete destruction." The psychological principle is that simple quantitative and factual descriptions do little to sway risk perception and by tapping into the emotional dimension of risk psychology through emotive terms, people are more likely to change their risk perceptions and act accordingly.

Sirens

During the height of the Cold War, Congress passed the Federal Civil Defense Act of 1950, which mandated the placement of outdoor sirens and public warning systems for use in case of nuclear attack (Coleman et al. 2011). Over time, political pressure prompted the adoption of these systems for use against natural hazards as well. By the early 1970s, many vulnerable communities began to utilize sirens for natural hazards (Coleman et al. 2011; Kessler 1970). In order to communicate warnings from the NWS to emergency managers (who activated the sirens) more quickly, the National Warning System (NAWAS) was developed. NAWAS facilitates warning communications across federal, state, and local emergency management. Sirens continue to be an important part of most integrated warning systems and have significant advantages over other systems. Sirens are an excellent way to warn those in the community that may not be otherwise aware of the developing hazard. Sirens cut across language barriers and are sensed by anyone outdoors. Given the long use of sirens, many people grew up associating sirens with a hazard, and instinctively expect to hear a siren prior to any immediate threat.

However, siren networks also have some significant disadvantages. Sirens are designed to only be heard outdoors. A typical siren is designed to produce at least

70 dB and to be heard at a range of at least 1 km; however, many residents may well live farther than 1 km from a siren or may be indoors away from an open window when the sirens are activated. Furthermore, the information conveyed by a siren is very limited; a siren basically warns the individual to seek shelter indoors, but cannot indicate the location or intensity of the hazard. Sirens may be operated by a third party, such as a local emergency manager, that may or may not have had adequate training; some false alarms may occur, or may sound too late for adequate shelter to occur. In addition, sirens are also used for a variety of other uses, such as by the local fire department to call-in volunteers. A national survey by the authors (Brotzge and Donner 2015) found that there is no standard procedure for when and under what circumstances sirens are to be sounded. While most restrict siren usage to tornado threats, some EMs sound sirens for tornadoes, hail, and wind. Finally, some metropolitan (and many rural) areas have no sirens largely due to their rather expensive costs for installation, operations, and maintenance.

Until relatively recently, sirens would sound for every tornado warning for your local county or city. But with the recent adoption of polygon shaped (storm-based) warnings, some siren systems can now be programmed to sound only when the tornado warning polygon overlaps the siren. In other words, digitized polygon warnings have helped reduce the number of siren false alarms, which in turn should make siren alerts more trusted.

Radio and Television

The invention and commercialization of radio by Guglielmo Marconi (1874–1937) in 1895, and later the electronic television in 1927 by Philo Farnsworth (1906–1971), would herald revolutionary changes to how weather observations were collected and how warnings were disseminated. Radio was first used by the Weather Bureau in 1905 to begin collecting observations from ships at sea, and then later from across Alaska (in 1907) (Calvert 1923). Simultaneously, radio was used to broadcast analyzed information back to those same areas. Daily radio weathercasts were begun in 1914, and special hurricane radio broadcasts were started in 1920. The radio expanded the speed, volume, and distance that information could be gathered and disseminated, which made a significant impact on maritime and aviation services.

The television brought even greater changes. The first television news weathercasts were broadcasted in the early 1950s. Some of these early weathercasts must have been rather rough; a presentation to the American Meteorological Society (AMS) in 1955 was titled, "The Challenge of Television Weather Programs" (Jehn 1956). The presentation by K. Jehn laid out eight separate actions approved by the Council of the AMS which aimed at improving the professionalization of on-air TV meteorology. This presentation was even reprinted as an article in the monthly journal of the AMS (Jehn 1956). To further these efforts, the AMS started the "AMS Seal of Approval" certification in 1957.

The dissemination of weather alerts was a natural part of TV weather from its inception. "Cut-ins" were used to interrupt normal programming with weather alerts; in the 1950s, cut-ins could be simply the text of the warning or the Emergency Broadcast System (EBS) tone and graphic (Coleman et al. 2011). By the mid-1960s, lengthier cut-ins became more common, with a meteorologist describing the hazard in more detail. The first live video of a tornado and the first radar hook echo shown on TV were broadcasted from the same station in Wichita Falls, TX, in 1964 (Coleman et al. 2011). TV meteorology continued to develop through the 1980s, as "crawls" and "bugs" were introduced (Coleman et al. 2011). Crawls were text that scrolled across the bottom of the screen, and bugs were icons usually placed near the top of the screen that highlighted that a particular warning was in effect. Later, more informative county maps were displayed, often color-coded to show the types of warnings that were in effect. As cable TV developed, whole stations were soon devoted to weather, and many local stations would have a secondary cable channel with a rotating loop of weather forecasts or radar image.

NOAA Weather Radio

The first NWS-sponsored radio stations began in the 1950s in order to communicate weather information more effectively for aviation applications (Coleman et al. 2011). By the 1960s, a network of 60 stations was providing forecasts and warnings to a wider variety of users, including aviation, maritime, and the general public. The first publicly available receivers were sold in 1968. About 125 stations were operating by 1974. Nevertheless, a post-event survey assessment following the April 1974 Super Outbreak found a significant need for improved warning dissemination across much of the country. Following that report, Congress appropriated funds to outfit 70% of the country with access to NOAA Weather Radio All Hazards (NWR), bringing the network total to over 300 stations (Coleman et al. 2011). These stations operate at wavelengths beyond AM and FM that require special receivers, and so these radios must be bought specifically for weather use.

A major advantage of NWR is that it provides continuous weather information directly to your home. More importantly, NWR is designed to sound an alert or tone whenever NWS warnings are issued. With only a few exceptions, nearly every county in the U.S. is now covered by NWR radio (Miller 2018). One disadvantage of NWRs was the fact that alerts would often sound for hazards that were irrelevant to your location; for example, a flood warning may not apply to you if you lived in a high location, or if the warning was for a different part of the county. This often led people to simply turn their radios off unless they knew of a hazardous situation ahead of time. A survey by Redmond (1995) found that only 24.6% of residents in an area of west Tennessee owned a NWR, and only one-third of those left it on continuously (Redmond 1995). Fortunately, in 1998 NWRs began utilizing Specific Area Message Encoding (SAME) (Miller 2018). This allows NWR users to limit the alerts to the county or sub-county of interest. In addition, an FCC rule was changed in 2002 to allow the user to select which types of warnings trigger the alert. These customizable features significantly reduce the number of false alarms and should improve the frequency of use of the system. To evaluate the effectiveness of NWR, Miller (2018) quantified the differences in tornado fatality and injury rates between those areas with and without NWR transmitters between 1970 and 2014. Miller found a nearly 40% reduction in injuries and a 10–50% reduction in fatalities for those areas that had NWR coverage.

A Decline in Tornadic Mortality and the Need for an Increased Focus on Social Science

In the present era, the technological innovations so described have resulted in a gradual reduction in severe weather mortality, a decline which began in 1925 which has steadily persisted in the decades to follow (Brooks and Doswell 2002). This was made possible from the first telegraph networks and weather vanes to modern remote sensing technologies, advances in weather modeling and AI, improvements in communications techniques and infrastructure, and the development of spotter networks, as well as the talent and commitment of NOAA personnel and forecasters and local, state, and county emergency managers. We have indeed come a long way. Yet, as Stokoe (2016) notes, the rate at which severe weather mortality is declining has decreased somewhat over the past three decades, a pattern which others have drawn attention to in earlier years (Ashley et al. 2008).

This may be because technology, however advanced, has reached a point of diminishing returns. This may be for a variety of reasons, all of which occur beyond the scope of the warning system itself and accordingly shape response. For instance:

1. *Warnings increasingly take place within a group context or naturally involve social interaction*

   Only in rare circumstances do warning recipients navigate the warning process alone. Warning response is a social event that involves periods of interaction with other individuals, groups, institutions, and organizations. Recipients may receive warnings while watching television as a family or in the presence of co-workers or interact with neighbors in the milling process of coming to terms with the ambiguity of weather alerts. For example, Kirschenbaum (1992) observed that two-thirds of victims of a gas explosion in Israel attempted to confirm warning information, a process repeatedly confirmed in the literature for decades (Drabek 1969). The social character of warnings is likely reinforced by the ubiquity of social media.
2. *Warning response is firmly tied to the socialization and social learning process*

   Previous experiences with disasters as well as our interactions with others shape how we behave and what we believe. For instance, if a community believes that mountains or rivers protect them from tornadoes, we are likely to be socialized into this belief (Donner et al. 2012). Also, if we experienced a number of false alarms in the past, we may be less likely to respond to future alarms. Much as we are socialized into education, work, and other institutions, so too are we socialized into natural hazard cultures.

3. *Whether persistent or situational, social norms form the basis for decision-making*

   The warning process does not postpone or suspend the dominant norms in place at the time when the warning is issued. By definition, a warning is intended to temporarily break the normative behavior of recipients. Indeed, conventional social norms are deeply resistant to the warning process as psychologically people hold on to what has come to be known as a "normalcy" bias (Okabe and Makami 1981). Such cognitive biases work to retain a sense of normality in the face of warnings, thereby neutralizing the risk communicated in the warning. At the same time, panic, which might be formally defined as a state of normlessness, has been long dispelled in the field (Quarantelli 1954). Thus, at any stage of the process, there is either some pre-existing or emergent behavioral code on which people rely to negotiate the risk implied in the warning.

4. *Socio-structural factors shape response*

   Finally, the hierarchical nature of society creates a structure in which some groups hold more resources than other groups. The poor (Fothergill and Peek 2004), racial and ethnic minorities (Fothergill et al. 1999), the elderly (Klinenberg 2003), and children (Peek et al. 2018; Peek 2008), for instance, all bear heavier risks from disaster for a variety of reasons, many of which involve poverty, lack of institutional trust, or disability. This may manifest itself, for instance, in difficulty affording a shelter, refusing to seek refuge in a public shelter, or inability reaching a shelter due to poor mobility.

Conclusion

The previous sections provide a general, historical overview of meteorology, locating the origins of modern warnings within a complex arrangement of scientific and social developments. Indeed, nearly 200 years of meteorological research have transformed the field, creating a largely successful, end-to-end warning model that has been emulated across the world. Historically, warnings have passed through a series of stages of development. Early warning systems were rooted in the crude technologies of the times, and, accordingly, rarely reliable. As time passed, advances in physics, chemistry, and the geosciences led to advances in the detection and prediction of these physical hazards, which, in turn, led to warning systems that were more effective in mitigating loss of life. Today, more than ever, we are aware of weather risks. At the same time, parallel social science research on the human dimensions of warnings—how people perceive, understand, and respond to warnings—has yielded important theoretical insights and policy solutions, but has in many ways lagged behind in theoretical development and empirical refinement.

The preceding historical account of meteorology offers an argument with respect to the overall need to further explore the development of sustained interdisciplinary research between the social and weather sciences. As it happens, without an understanding of human psychology and social behavior, policymakers must draw from intuition or common belief, neither of which offer reliable accounts of human behavior, and therefore no dependable foundation for policy. The panic myth, which served as a foundation for warning policy until as late as the 1950s, presents a stark example in this regard. Presumed panic in the face of disaster, a long-dispelled myth in the social science literature, cast an enduring, decades-long suspicion over warnings. As a result, research into warning systems faltered until a decade of unprecedented severe weather compelled the weather community to reconsider the viability of severe weather forecasting. With panic now regarded as rare, warning systems have seen immense growth in scope, complexity, and use. Of course, early meteorology cannot be blamed, for social science, and especially disaster research, was largely in its infancy at the time. However, by ensuring collaboration between the social sciences and the natural sciences, we can now better refine and shape our warning policies to much greater effectiveness.

The question therefore remains: how do we synthesize these two perspectives—natural science and social science—into a sustainable, methodological, and reliable program in order to improve the effectiveness of warning systems? Towards this end, Chaps. 3 and 4 provide a detailed literature review of our present understanding of the warning response model.

Bibliography

Abbe, C. (1906). Benjamin Franklin as Meteorologist. *Proceedings of the American Philosophical Society*, 183(45), 117–128.

Agnew, D. (2002). History of Seismology. *International Geophysics*, 81, 3–11. doi: https://doi.org/10.1016/S0074-6142(02)80203-0.

Ashley, W. S., Krmenec, A. J., & Schwantes, R. (2008). Vulnerability due to Nocturnal Tornadoes, Weather and Forecasting, 23(5), 795-807.

Aspaas, P. and Hansen, T. (2012). The Role of the Societas Meteorologica Palatina (1781–1792) in the History of Auroral Research. *Acta* Borealia, 29(2), 157–76. doi: https://doi.org/10.1080/08003831.2012.732283.

Bradford, M. (1999). Historical Roots of Modern Tornado Forecasts and Warnings. *Weather and Forecasting, 14*(4), 484–91. doi: https://doi.org/10.1175/1520-0434(1999)014<0484:HROMTF>2.0.CO;2.

Brooks, H. and Doswell C. (2002). Deaths in the 3 May 1999 Oklahoma City Tornado from a Historical Perspective. *Weather and Forecasting*. 17(3): 354–361.

Brotzge, J. and Donner W. (2015). The general use of outdoor warning sirens: a preliminary survey of emergency managers. *Journal of Emergency Management*, 13(1), 61–69.

Calvert, E. B. (1923). History of Radio in relation to the work of the Westher Bureau. *Monthly Weather Review,* 51(1), 1–2. doi: https://doi.org/10.1175/1520-0493(1923)51<1:HORIRT>2.0.CO;2.

Coleman, T.A., Knupp, K.R., Spann, J., Elliott, J.B., and Peters, B. (2011). The History (and Future) of Tornado Warning Dissemination in the United States. *Bulletin of the American Meteorological Society*, 92(5), 567–82. doi: https://doi.org/10.1175/2010BAMS3062.1.

Coleman, T. A., and Pence, K. (2009). The Proposed 1883 Holden Tornado Warning System: Its Genius and Its Applications Today. *Bulletin of the American Meteorological Society*, 90(12), 1789–96. doi: https://doi.org/10.1175/2009BAMS2886.1.

Donner, W.R., and Lavariega-Montforti, J. (2018). Ethnicity, Income, and Disaster Preparedness in Deep South Texas, United States, *Disasters* 42(4), 719–33. doi: https://doi.org/10.1111/disa.12277.

Donner, W.R., Rodriguez, H, and Diaz, W. (2012). Tornado Warnings in Three Southern States: A Qualitative Analysis of Public Response Patterns. *Journal of Homeland Security and Emergency Management*, 9(2), 1–21.

Doswell, C.A., Moller, A.R., and Brooks, H.E. (1999). Storm Spotting and Public Awareness since the First Tornado Forecasts of 1948. *Weather and Forecasting*, 14(4), 544–57. doi: https://doi.org/10.1175/1520-0434(1999)014<0544:SSAPAS>2.0.CO;2.

Drabek, T.E. (1969). Social Processes in Disaster: Family Evacuation. *Social Problems*, 16(3), 336–49. doi: https://doi.org/10.2307/799667.

Entrekin, H.D., Wilson, J.W., and Hage, K.D. (1969). Evaluation of the Atlantic City Mesonet for Short Range Prediction of Aviation Terminal Weather. *Journal of Applied Meteorology and Climatology*, 8(4), 473–83. doi: https://doi.org/10.1175/1520-0450(1969)008<0473:EOTACM>2.0.CO;2.

Fleming, J.R. (1990). *Meteorology in America, 1800-1870*. Baltimore, MD: Johns Hopkins University, Baltimore, MD.

Fothergill, A. and Peek, L. (2004). Poverty and Disasters in the United States: A Review of Recent Sociological Findings. *Natural Hazards*, 32, 89–110. doi: https://doi.org/10.1023/B:NHAZ.0000026792.76181.d9.

Fothergill, A, Maestas, E., and Derouen, J. (1999). Race, Ethnicity and Disasters in the United States: A Review of the Literature. *Disasters*, 23, 156–73. doi: https://doi.org/10.1111/1467-7717.00111.

Gillan, J. n.d. "Ancient Chinese Earthquake Detector Invented 2,000 Years Ago Really Worked!" Retrieved December 27, 2020 (https://www.ancient-origins.net/ancient-technology/incredible-earthquake-detector-invented-nearly-2000-years-ago-001377).

Glaser, R and Riemann, D. (2009). A Thousand-Year Record of Temperature Variations for Germany and Central Europe Based on Documentary Data. *Journal of Quaternary Science*, 24(5), 437–49. doi: https://doi.org/10.1002/jqs.1302.

Grazulis, T. P. (2001). *The Tornado: Nature's Ultimate Windstorm*. Norman, OK: University of Oklahoma Press.

Grice, C.K., Trapp, R., Corfidi, S., Davies-Jones, R., Buananno, C., Craven, J., Droegemeier, K., Duchon, C., Houghton, J., Prentice, R., Romine, G., Schlachter, K., and Wagner, K. (1999). "The Golden Anniversary Celebration of the First Tornado Forecast." *Bulletin of the American Meteorological Society* 80(7):1341–48. doi: https://doi.org/10.1175/1520-0477(1999)080<1341:TGACOT>2.0.CO;2.

Hellmann, G. (1908). The Dawn of Meteorology. *Quarterly Journal of the Royal Meteorological Society*, 34(148), 221–32. doi: https://doi.org/10.1002/qj.49703414802.

Huang S-K, Lindell MK, and Prater CS. 2016. Who Leaves and Who Stays? A Review and Statistical Meta-Analysis of Hurricane Evacuation Studies. *Environment and Behavior*, 48(8), 991–1029. doi: https://doi.org/10.1177/0013916515578485

Hughes, P. (1970). *A Century of Weather Service: A History of the Birth and Growth of the National Weather Service, 1870-1970*. Gordon and Breach.

Jackson, M. (2011). Raindrops over Joseon – Sejong's Cheugugi | London Korean Links.". https://londonkoreanlinks.net/2011/10/22/raindrops-over-choseon-sejongs-cheugugi/. Retrieved December 27, 2020.

Jehn, K. H. (1956). The Challenge of Television Weather Programs. *Bulletin of the American Meteorological Society*, 37(7), 351–53. doi: https://doi.org/10.1175/1520-0477-37.7.351.

Kessler, E. (1970). Tornadoes. Bulletin of the American Meteorological Society, 51(10), 926–936.

Kirschenbaum, A. (1992). Warning and Evacuation During a Mass Disaster: A Multivariate Decision-Making Model. *International Journal of Mass Emergencies and Disasters*. 10, 91-114.

Klinenberg, E. (2003). *Heat Wave: A Social Autopsy of Disaster in Chicago*. 1st Edition. Chicago London: University Of Chicago Press.

Landsberg, H. E. (1964). "Early Stages of Climatology in the United States." *Bulletin of the American Meteorological Society* 45(5):268–75. doi: https://doi.org/10.1175/1520-0477-45.5.268.

Leik, R.K., Carter, T.M., Clark, J., Kendall, S. and Gifford, G. (1981). *Community Response to Natural Hazard Warnings*. Minnesota University Minneapolis.

Lindell, M. K., Huang, S., Wei, H., and Samuelson, C. (2016). "Perceptions and Expected Immediate Reactions to Tornado Warning Polygons." *Natural Hazards* 80(1):683–707. doi: https://doi.org/10.1007/s11069-015-1990-5.

Lindell, M.K., and Perry, R.W. (1992). *Behavioral Foundations of Community Emergency Planning*. Washington, DC, US: Hemisphere Publishing Corp.

Mahmood, R., Boyles, R., Brinson, K., Fiebrich, C., Foster, S., Hubbard, K., Robinson, D., Andresen, J., and Leathers, D. (2017). Mesonets: Mesoscale Weather and Climate Observations for the United States. *Bulletin of the American Meteorological Society*, 98(7), 1349–61. doi: https://doi.org/10.1175/BAMS-D-15-00258.1.

Mileti, D. S., and Sorensen, J.H. (1990). *Communication of Emergency Public Warnings: A Social Science Perspective and State-of-the-Art Assessment*. ORNL-6609. Oak Ridge National Lab., TN.

Mileti, D. 1999. *Disasters by Design: A Reassessment of Natural Hazards in the United States*. Washington, D.C.: Joseph Henry Press.

Miller, B.M. (2018). The Not-So-Marginal Value of Weather Warning Systems. *Weather, Climate, and Society*, 10(1), 89–101. doi: https://doi.org/10.1175/WCAS-D-16-0093.1.

MIT. n.d. John Milne. https://lemelson.mit.edu/resources/john-milne. Accessed 8/18/21.

Moore, P. (2015). The Birth of the Weather Forecast - BBC News. https://www.bbc.com/news/magazine-32483678. Retrieved December 28, 2020.

Morss, R.E., Mulder, K., Lazo, J.K. and Demuth, J.L. (2016). How Do People Perceive, Understand, and Anticipate Responding to Flash Flood Risks and Warnings? Results from a Public Survey in Boulder, Colorado, USA." *Journal of Hydrology*, 541, 649–64. doi: https://doi.org/10.1016/j.jhydrol.2015.11.047.

NASA. 2002. Towards Numerical Weather Prediction. Accessed 6/20/2022 from https://earthobservatory.nasa.gov/features/WxForecasting/wx3.php.

National Cancer Institute. "Changing Adolescent Smoking Prevalence". Tobacco Control Monograph No. 14. Bethesda, MD: U.S. Department of Health and Human Services, National Institutes of Health, National Cancer Institute. NIH Pub. No. 02-5086, November 2001.

NOAA. n.d. "History of the National Weather Service." https://www.weather.gov/timeline. Accessed August 1, 2019.

NWS. n.d. History of the National Weather Service. Accessed 6/20/2022 from https://www.weather.gov/timeline.

Okabe, K and Makami, S. (1981). A Study on the Socio-Psychological Effect of A False Warning of the Tokai Earthquake in Japan. Paper presented at the 10th World Congress of Sociology in Mexico City, August 22, 1982.

Peacock, W., Gladwin, H. and Morrow, B (eds.) (1997). *Hurricane Andrew: Ethnicity, Gender and the Sociology of Disasters*. 1st edition. London: Routledge.

Peek L., Abramson D.M., Cox R.S., Fothergill A., Tobin J. (2018) Children and Disasters. In: Rodríguez H., Donner W., Trainor J. (eds) Handbook of Disaster Research. Handbooks of

Sociology and Social Research. Springer, Cham. https://doi.org/10.1007/978-3-319-63254-4_13

Peek, L. (2008). Children and Disasters: Understanding Vulnerability, Developing Capacities, and Promoting Resilience — An Introduction. *Children, Youth and Environments*, 18(1), 1–29.

Perry, R.W. and Lindell, M.K. (1986). "Twentieth-Century Volcanicity at Mt. St. Helens: The Routinization of Life Near an Active Volcano." Tempe, AZ: School of Public Affairs, Arizona State University.

Powell, S. (2008). Communicating Weather Information to the Public: People's Reactions and Understandings of Weather Information and Terminology. 3rd Symposium on Policy and Socioeconomic Impacts, AMS Annual Meeting, New Orleans, LA.

Quarantelli, E. L. (1954). The Nature and Conditions of Panic. *American Journal of Sociology*, 60(3), 267–75. doi: https://doi.org/10.1086/221536.

Redmond, J.W. (1995). NOAA Weather Radio as an Emergency Communication Vehicle in West Tennessee. *Weather and Forecasting*, 10(3), 485–97. doi: https://doi.org/10.1175/1520-0434(1995)010<0485:NWRAAE>2.0.CO;2.

Rodríguez, H., Donner, W., and Trainor, J.E. (eds.) (2017). *Handbook of Disaster Research*. 2nd ed. 2018 edition. New York, NY: Springer.

Sherman-Morris, K. (2005). "Tornadoes, Television and Trust—A Closer Look at the Influence of the Local Weathercaster during Severe Weather." *Global Environmental Change Part B: Environmental Hazards*, 6(4), 201–10. doi: https://doi.org/10.1016/j.hazards.2006.10.002.

Simmons, K.M., and D. Sutter. 2009. "False Alarms, Tornado Warnings, and Tornado Casualties." Weather, Climate and Society, 1: 38-53.

Smithsonian Institute. 2017. Joseph Henry: The Father of Weather Forecasting (And the First Smithsonian Secretary!)." Smithsonian Science Education Center. Retrieved December 28, 2020 (https://ssec.si.edu/stemvisions-blog/joseph-henry-father-weather-forecasting-and-first-smithsonian-secretary).

Sorbjan, Z. (1996). *Hands on Meteorology: Stories, Theories, and Simple Experiments*. 1st edition. Washington, D.C.: American Meteorological Society.

Spilhaus, AF. (1950). Progress in Meteorological Instrumentation, 1920–1950. *Bulletin of the American Meteorological Society*, 31(10), 358–64. doi: https://doi.org/10.1175/1520-0477-31.10.358.

Stokoe, R.M. (2016). "Putting People at the Centre of Tornado Warnings: How Perception Analysis Can Cut Fatalities." International Journal of Disaster Risk Reduction, 17, 137–53. doi: https://doi.org/10.1016/j.ijdrr.2016.04.004.

Sutter, D., & Erickson, S. (2010). The Time Cost of Tornado Warnings and the Savings with Storm-Based Warnings, Weather, Climate, and Society, 2(2), 103-112.

US Department of Commerce, NOAA n.d.-a. "The History of Numerical Weather Prediction." (https://celebrating200years.noaa.gov/foundations/numerical_wx_pred/welcome.html#early).

US Department of Commerce, NOAA. n.d.-b "Cooperative Observer Program (COOP)." (https://www.weather.gov/coop/overview).

US Department of Commerce, NOAA. n.d.-c "SKYWARN." (https://www.weather.gov/skywarn/).

US Geological Survey. "Education." Retrieved December 29, 2020. (https://www.usgs.gov/natural-hazards/earthquake-hazards/education#eighteenth).

Westfall, R.S. (1985). Science and Patronage: Galileo and the Telescope. *Isis*, 76(1), 11–30.

Whitnah, D.R. (1961). *A History of the United States Weather Bureau*. 1st edition. Urbana-Champaign, IL: University of Illinois Press.

Williamson, F. (2015). Weathering the empire: Meteorological research in the early nineteenth-century Straits Settlements nineteenth-century Straits Settlements. *The British Journal for the History of Science*, 48 (3), 475–492.

Chapter 3
Public Warnings Response

Following nearly four decades of social science warning research, scholars have come to identify and appreciate more fully the social forces that oversee how people and communities receive, interpret, understand, and respond to warnings. Our scientific attitudes towards the warning process have fundamentally evolved over this period. We better grasp, for instance, the role of false alarms in producing skepticism, the elements of warning messages that tend to cause confusion, and the social factors that are known to either aid or obstruct the final act of response. Warning science has also served to dispel dangerous misconceptions about panic that had for so long prohibited the use of tornado warnings. Without question, we know far more about warning response today than we did in the past. Lacking, however, is a general framework or theory for organizing the literature on warnings, one that would provide a useful context for interdisciplinary research and meteorological application. We resolve the problem by applying a modified version of Mileti and colleagues' stage-based model to organize what is at once a useful but amorphous body of scholarship on warning response into an intuitive framework useful for scientists, policymakers, and emergency managers.

The chapter is composed of two sections. The first section opens with a brief overview of Mileti and colleagues' stage-based model, followed by a summary of empirical research on warnings at each stage of the model. For instance, in our treatment of the "belief" stage of the model, we discuss at length the literature on factors that govern trust in local authorities, the psychological effects of false alarms on warning credibility, and the critical role of environmental cues in tipping the scales from skepticism to belief. The goal is to articulate in as much detail as possible the factors, behaviors, and events that shape what occurs at each stage of the warning process. As discussed, the stages are (a) *reception*, (b) *understanding*, (c) *belief*, (d) *confirmation*, (e) *personalization*, (f) *believe protective action feasible*, and (g) *protective action*. This chapter also notes the importance of resource availability, access and knowledge of the warning response process. To this are added several emergent concepts within the literature that also govern and shape warning response:

W. Donner et al., *Severe Weather Warnings: An Interdisciplinary Approach*,
https://doi.org/10.1007/978-3-031-05031-2_3

(a) *resource availability*, (b) *knowledge of resources*, (c) *accessing resources* (d) *knowledge of how to use resources*.

The second section explores a far less investigated element of the warning process: *how* the transitions between model stages occur—namely, the dynamic features of the model. Although we understand well what happens *within* each model stage (as will be demonstrated in the first section of the chapter), the psychological and social undercurrents that steer stage transitions remain by-and-large uncharted. Chief among these dynamics is whether individuals pass through model stages in a linear fashion—that is directly from stage to stage, uninterrupted from reception through protective action—or whether there exists recursive aspects to the model, meaning that individuals may "loop back" to earlier stages under certain circumstances. The three dynamic elements of the model explored here are *interdependence* (i.e., how earlier decisions affect later decisions), the *directionality of the model* (i.e., does it uniformly flow in one direction), and the *inconsistency of demographic predictors* across multiple stages. Indeed, as will be shown, warning response models cannot escape a deep complexity.

Warning Response Model

Mileti and Sorensen (1990) characterize warnings as a complex system itself composed of three sub-systems: (1) The *detection subsystem*, which involves the monitoring, data collection, and evaluation of the environment (e.g., scientists, NWS, SPC, forecasters); (2) the *management subsystem*, involving the interpretation, application, and coordination of information (e.g., emergency managers); and (3) the *response subsystem*, encompassing the processes by which the community receives, understands, and responds to information (this chapter deals mainly with the *response* subsystem, whereas Chap. 4 deals with the *detection* and *management* subsystems).

Initial work suggested that the public passes through six distinct stages when responding to a warning: (1) hearing, (2) understanding, (3) believing, (4) personalizing, (5) deciding and responding, and, finally (6) confirming the warning. This model would be expanded in later scholarship. In his book, *Disasters by Design*, Mileti (1999) adapted Lindell and Perry's (1992) response model to prior research by Mileti and Sorensen (1990), resulting in the full model presented under Fig. 3.1. According to this synthesized model, there are roughly eight stages to the process by which people evaluate and respond to warnings: (a) reception, (b) understanding, (c) belief, (d) confirmation, (e) personalization, (f) assessment of available resources, (g) access to available resources, and h) protective action (See Fig. 3.1). Thus, scholars have increasingly come to regard warning response as not merely a single act, but rather a set of discrete, progressive steps through which people traverse either individually or more frequently, collectively. If at any stage there is a "failure", then the individual will either not respond or will respond in a less-than-ideal manner. The following section further develops the model in detail, offering

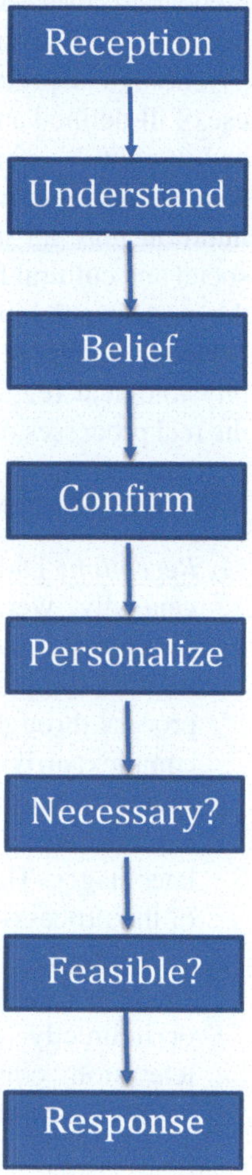

Fig. 3.1 Visualization of the warning response process (Mileti and Sorensen 1990; Lindell and Perry 1992; Mileti 1999)

additional clarification and propositions at each stage, and offers further conceptual and analytic clarification regarding processes and outcomes.

Warning Response in the Social System

While a great deal of focus and attention has been paid to the scientific and emergency subsystems, comparatively less attention has been devoted to understanding how warnings operate within social systems. How, then, might we begin to organize our understanding about how warnings operate within the social system?

Fortunately, the model of warning response, described earlier, offers a means by which to structure our understanding of response within the social system. The model describes a series of universal behavioral and thought processes that occur in response to risk and is useful as it avoids reducing warning response to mere psychological processes or states. Furthermore, this model moves us beyond the use of ill-defined and empirically-weak explanations related to "instinct," "impulse," or "stupidity"—none of which provide a strong theoretical basis for illuminating warning response in sociological and scientific terms. While cognition, emotion, and intuition play an important role, the model also acknowledges the influences of social and cultural factors on behavior and decision making during periods of risk. In the model, social interaction, social norms and beliefs empirically influence warning behavior and play more prominent roles alongside psychological (e.g., fear) or physiological (e.g., stress) influences. The result is a model of greater fidelity to the real processes of warning response. The reconceptualization of warning response as a process stands as one of the most important innovations in the field of disaster sociology. Here we elaborate on the model further.

(a) *Reception*: Did the person at any point receive the official warning or, more generally, were they made aware of the risks of imminent severe weather? *Reception* is defined as the first contact with information regarding a severe weather event already occurring or predicted to occur within the near future. The process through which the community receives warning information is deeply complex, arriving through any number of channels, from any number of sources, and with varying degrees of reliability and accuracy. How and under what circumstances warning messages are received may have profound effects on later stages. There are three dimensions to reception that may affect later stages of the process—the (a) medium, (b) origin, and (c) content.

Medium—how the recipient received the warning message. Recipients may encounter a warning visually or audibly through some formal institution, directly or indirectly. As examples, one might observe an NWS-issued crawler on television; come across a warning on weather.com; or hear the local siren network. In practical terms, the medium is important, as we know that access to warning information varies widely across social groups and that some social groups are more or less likely to receive official messages than others. For instance, during Hurricane Andrew, Hispanics relied much more heavily on storm information sources outside of formal channels (Peacock et al. 1997). Informal networks are a critical component of this process and may impact the accuracy and reliability of the information groups receive.

Origin—the organization from which the message originated. Or perhaps more importantly, the organization from which the information is perceived to have originated. This is significant because we know that recipients hold different levels of trust from different institutions. Members of the public, for instance, notably develop trust in certain news organizations over time, thereby affecting what they regard as credible (Sherman-Morris 2005). Thus, even if two stations hold the same information, the station with which the recipient does not hold a relationship of trust may be taken less seriously.

Content—severe weather information may come in a variety of different forms. It may come in the form of a warning, a watch, an advisory, or an outlook, each of which expresses different levels of risk. This also impacts response as people assess and personalize these events based on their understanding (or lack thereof) of what these concepts actually mean, or they simply do not fully comprehend these concepts and the connotations of the immediacy of danger.

While warnings may arrive in a wide variety of forms and from a wide variety of sources, the most crucial element is whether imminent *risk or danger was communicated by them and received in some form.*

(b) <u>Understanding</u>: Does the recipient comprehend the meanings contained within the message? Within the context of understanding warnings, recipients may ask themselves, Who issued the warning? In what language is it conveyed? What is the difference between a tornado "outlook," "warning," and "watch?" What is a "short-term forecast" as opposed to a "hazardous weather outlook?" To understand a warning signal or alert is not merely to grasp the implied risk, but also the recommendations and commands stated therein. Thus, warnings communicate two things: (a) the level of risk the recipient is estimated to face and (b) what should be done to minimize the estimated risk. Sometimes both dimensions are communicated simultaneously within the same medium. NWS warnings, for instance, communicate whether a warning or watch is issued and what to do (e.g., shelter in place). On the other hand, separate documents or sources may act as a risk alert, whereas a separate document or source may provide information on how to reduce risk. In the latter case, for instance, during hurricanes NOAA typically issues watches and warnings whereas a vast majority of the information regarding evacuation and safety comes from local governments. It should be further emphasized that risk comprehension is not simple and varies widely across cultures. For example, the literature shows that recipients oftentimes do not grasp the general differences between watches and warnings. In a study of the public's use of weather information, Powell (2008) found that nearly one third of the respondents could not distinguish between watches and warnings, and more recent research demonstrates that confusion surrounds flash flood warnings, as well (Morss et al. 2016).

At this stage, misunderstanding a warning yields several different outcomes for the aforementioned model.

(i) The recipient understands the risk to be *lower* than implied in the message, thus leading to perceptions of risk lower than intended, or leading to inaction or action below the required threshold for protection.
(ii) The recipient understands the risk to be *higher* than implied in the message, thus leading to perceptions of risk that are greater than intended and leading to action that is possibly unnecessary. This may in turn lead to a higher perception of false alarms as, all else equal, the eventual risk does not equal the expected risk.

Warnings often also direct the recipient in terms of what to do, i.e., "seek shelter immediately!" People without proficiency in the language through which warnings are communicated may not entirely grasp these directions or the subtle meanings implied.

(c) Belief: Does the recipient believe that the level of risk suggested in the message reasonably approximates the state of the environment (this is not to be confused with a later stage, risk personalization, in which the recipient regards the risk, which he or she judged as credible in the general sense, as one that is personally threatening or not)? At this stage, the recipient of the warning asks, "is the message credible enough to persuade me that severe weather has developed—not necessarily close to me or close enough to start shelter preparation—but at the very least *somewhere* close enough to be threatening." At this point, a dynamic range of factors that are internal and external to the recipient shape this belief. External factors may include the coherence of the message and whether the source of such messages are thought credible (Mileti and Sorensen 1990). Factors in the immediate environment also play a role. Interactions with others, or evaluating how others are reacting to the warning, shape belief in the likelihood of severe weather. Friends' and neighbors' response—or the lack thereof—define circumstances as normal or worthy of concern for the recipient (Dash and Gladwin 2007). Internal factors include a broad range of psychological factors shaping credulity. Past experience with disasters, trust in institutions (Donner et al. 2012), and encounters with false alarms (to be discussed later, Simmons and Sutter 2009) mold the innermost perception of risk among those having advanced to the belief stage. The process is a complicated one and, much like other predictive models in the social sciences, e.g. voting behavior, the etiology of risk belief is deeply complex, multifaceted, and dynamic.

Making matters more complicated, recipients may either believe or disbelieve the *understood* level of risk. Thus, recipients may believe an incorrectly understood message or disbelieve a correctly understood message (see previous section). Ideally, they should believe a correctly understood message, if by "correctly understood" we mean an understanding that aligns with the meaning intended by the group or institution from which the message originated. *It is important to note that just because a recipient is progressing through the model does not mean such progress necessarily improves safety.* We will further discuss this important point later.

(d) Confirmation, i.e., recipients check the reliability and validity of the information they receive. Scholars consistently note the striking consistency with which confirmation behavior is observed (Huang et al. 2016). Indeed, there is perhaps no stage more well documented than that of confirmation; it is well known that recipients rarely act immediately or exclusively on a single piece of information. Rather, they normally collect, or attempt to collect, multiple sources of information and evaluate them collectively to develop an overall assessment of the situation. What is unique about the confirmation stage, and what is unique in how we regard it from past applications, is that confirmation may take place at nearly any stage. Recipients do not merely confirm beliefs, for instance, but also confirm whether their understanding of information is correct (discussed earlier) or whether, once risk is personalized, the correct protective action is being taken (discussed later).

In practice, confirmation may assume a variety of forms. Recipients may casually check outside to detect darkening clouds or thunder, or may call any number of people or persons to discuss a recent severe weather watch. The impulse to confirm is persistent, even information that arrives from formal sources otherwise regarded as reliable and legitimate will likely be subject to confirmation. For instance, NWS watches and warnings are frequently confirmed with informal sources, such as friends and family. Recipients evaluate information for consistency and as a result, sheltering or evacuation is oftentimes delayed.

(e) *Personalization*: The recipient believes a hazard risk is imminent, explicitly real, and poses an unambiguous threat to self or family. Herein lies a distinction with stage c ("belief"). For stage c, the recipient adopts a belief that somewhere in the environment exists a risk or at the very least, a risk of such likelihood that confirmation initiates. In abstract terms, there is a risk "out there" somewhere. It remains unclear, however, whether the risk in question is a personal threat, a threat to someone else, or whether the risk is sufficiently small as to be comfortably disregarded. Thus, stages (c) and (e), while both dealing with belief, are different in that (c) is belief set in general or abstract terms and stage (e) is in concrete, personal terms. Consider common mythologies surrounding severe weather that suggest that some towns are "immune" to tornado activity, mobile homes attract tornadoes, or that mountains and rivers afford protection from tornadoes (Grazulis 2001). These beliefs are myths and thus are not "real." Nevertheless, their consequences for the warning process are quite real. For instance, residents may not seek shelter because although they believe (stage c) that tornadoes are present, they do not personalize the risk because they hold a cultural view that a local river will block the tornado's passage.

(f) *Protective Action Necessary*: From the personalization stage, the recipient begins to evaluate the resources he or she has for survival and asks: will doing something make me safer? At this stage, recipients interpret danger as at least sufficient to begin consideration of possible courses of action to mitigate risks. Note that protective action has not yet begun. Any action or behavior at this stage is limited to the search for additional information to confirm knowledge about protective action. For instance, calling a co-worker to confirm whether seeking shelter under an overpass is advisable typically takes place within stage (f). The perceived absence of resources may halt the process even if resources are available or, alternatively, lend undue credibility to decisions unlikely to improve chances of survival. Stage (f) rests on an awareness of resources. Recipients, for example, may not know of a local shelter, or may decide that flight from an approaching tornado is advisable over sheltering in place despite the presence of a basement or shelter. Again, since tornado warnings tend to merge with the events of everyday life, recipients at this stage evaluate the costs and benefits of responding to a warning. Yes, a warning has been issued, but will I get into trouble at work if I leave? In many cases, institutions remain open even during flood warnings and tornado warnings, thereby lending a persistent sense of normality within the community. If local businesses, schools, and government are open, this implies that the risk is not too great, or at least not great enough to continue evaluating a response, and so responding may be regarded as *unnecessary*. Thus, individuals within the community must decide whether the benefits of seeking shelter outweigh the costs of violating normative obligations with the community.

(g) *Protective Action Feasible*, here recipients make use of resources or access resources identified in stage (f). Access to resources previously believed to be helpful in the response process may not be available for a variety of reasons. Recipients may be unable to physically access the resources or, alternatively, may not trust those who make the resource available to the public. For instance, undocumented migrants may be fearful of using public shelters for fear of deportation despite knowing that shelters offer the best protection under tornado warnings (Donner 2007). Prior preparations on the part of individuals and the community now factor heavily into the warning process (Donner and Lavariega-Montforti 2018). It therefore stands that some less-prepared individuals may have significantly more difficulty taking shelter than those who were able to prepare.

(h) *Protective Action*. Protective action occurs when a warning is received, understood, believed, and acted upon after identifying and obtaining the necessary resources to do so. Note that the model does not suggest that the protective action in question will necessarily, by virtue of merely occurring, improve safety. Rather, it only posits that the recipient has at this point deemed the risks sufficient to take action. Individuals or families living in poverty face well-known challenges in preparing for and evacuating from disasters (Fothergill and Peek 2004). Alternatively, people with disabilities or the elderly may suffer difficulties during egress from buildings or in complying with a plan of escape. Children may also need extensive assistance during the evacuation process (Peek 2008).

Model Stages: A Review of Literature

The following section provides a comprehensive "deep dive" into the stages of the warning response model. While the broad stages of the model are useful in reorienting our thinking about warning response, what is known to occur during each specific stage and how factors related to psychology, social factors, and culture affect their resolution has never been fully described. Here we review the complexities of each stage in detail in order to better understand the need for an interdisciplinary approach.

Reception

Recall the first stage of the model, *reception.* Does the warning message reach the intended recipient in some way? Some warnings are received, others missed, but the process by which warnings are received or missed is far from accidental. A variety of factors facilitate or inhibit the transmission of warning information to the individual, and, in broader terms, the diffusion of public warnings is rarely uniform across different populations, groups, and contexts. For instance, bilingual populations may fail to receive alerts, and older populations, with much less experience with technology as a general rule, may miss emergency warnings communicated through social media. What are some of the common factors that shape the reception of issued warnings?

Informal Social Networks

Despite being a well-documented feature of the disaster experience, the role of social networks in warning response has received regrettably little attention from scholars in recent years. This is despite the fact that we know informal networks play a significant role in the reception process. In sociological terms, social networks comprise the unofficial, normative relationships and bonds within the community. Friendships, family institutions, co-worker relationships, civic clubs and associations, and religious groups, among others, act as main hubs uniting the complex system of social arteries through which information circulates within a population. Customarily, people well-embedded within the established informal networks of a community are more likely to receive information of any kind. On the contrary, information communicated informally tends to bypass the more isolated, friendless, and those generally unwoven into the fabric of the community. As with other forms of information, warnings tend to follow the same set of rules when issued to the public: those connected to well-knit community networks are more likely to receive warnings, whereas those outside of such networks are less likely to receive warnings. Taken together, evidence suggests that informal social networks play a distinct and important role in the transmission of warning information, and community members lacking strong social networks are less likely to receive a formal warning in comparison to peers well integrated into local social networks (Scanlon and Frizzell 1979).

Kirschenbaum (1992) offers some of the earliest and most compelling empirical evidence for this reliance on social networks. Among victims of a gas explosion in Israel, a majority—roughly two-thirds— sought information through social networks: 45% spoke to immediate neighbors, 12.6% called police, 9.5% called others in the neighborhood, 13.7% left shelter to clarify what was happening, and 18.9% turned to the media. Related effects were witnessed after a flood event in Australia where most (55%) of respondents described receiving alerts initially via informal sources followed thereafter (40%) by formal sources (Molinari and Handmer 2011).

Indeed, there is no shortage of evidence pointing to the importance of social networks and its impact on the warning process. For example, message reception is shaped by the recurring interaction of family members (Lardry and Rogers 1982), robust social or network involvement (Rogers and Nehnevajsa 1987; Perry and Green 1982; Sorensen and Gersmehl 1980; Turner et al. 1979), and consistent relationships with a subculture or voluntary association (Perry et al. 1981). Given the critical relevance of social networks in the warning reception process, it is perhaps not surprising that a majority of those who perished in the Chicago heatwave of 1995 lived relatively solitary lives. Klinenberg (2001: 501) credits the fatalities to an "increase in the number and proportion of people living alone, including seniors who outlive or become estranged from their social networks." From this, we might also reasonably infer that recent immigrants, the homeless, and tourists, all of whom likely lack robust local network involvement, may indeed be less likely to receive warnings. It would be an error to consider social networks as plainly "filling in the gaps" passed over by formal warnings. Rather, informal warnings are a core element of the warning process and an important, primary channel of communication for many.

Some evidence suggests that networks among marginalized minority groups commonly emerge as an adaptive response to an exclusion from both the formal communication networks of a community (e.g. emergency management) as well as informal networks associated with majority groups. Donner (2007), found that local undocumented immigrants in a tornado-prone region in Tennessee relied heavily on local church institutions for information critical to response and sheltering. Oftentimes, particularly among Latino communities, trusted social institutions (e.g., local Hispanic-serving churches) act as representatives on behalf of undocumented immigrants who are unable or understandably unwilling to interact directly with local authorities. Thus, informal networks seem an almost inevitable feature of warning reception, so much so that if none exist, they will naturally form and evolve as a method of adaptation. This natural resilience should be harnessed by local officials as a means by which to reach isolated groups.

Language Barriers to Reception

It is well documented in the literature that language barriers directly and indirectly obstruct warning communication. Illustrating this point, Aguirre (1988) documented the failure on the part of government and media to address the linguistic needs of Mexican-Americans during the Saragosa, Texas, tornado outbreak on May 22, 1987. Such difficulties in warning reception are observed in other contexts. Following the Tuscaloosa 2011 tornado outbreak, similarly to Aguirre scholars again reported very low levels of threat awareness among Hispanics. Risk perception surveys suggested that only about one-third of Hispanic respondents were aware at some point that the threat existed compared to 90% of non-Hispanic whites (Senkbeil et al. 2014). What is more, even when recipients hold a basic familiarity with the language in which warnings are communicated, problems with understanding remain. This is because warnings provide more than a simple and unnuanced alert to a disaster. Rather, warnings outline the nature of the disaster, the location of the disaster, provide detailed advice on what to do, and so on. Non-native speakers may acquire the general sense of a warning, but nevertheless miss important details due to linguistic nuance and semantics.

Demographic Determinants of Reception

In general, our understanding of the demographics of reception is confined to only a few large-scale, representative studies (Ripberger et al. 2019). Thus, at present there remains challenges to advancing generalizations about the role of race, class, gender and other sociodemographic factors in the reception process. With that in mind, several studies nevertheless demonstrate that demographics do indeed shape whether or not one receives, hears, or witnesses a warning either through formal (e.g., NWS) or informal (e.g. friends) channels of communication. Perhaps the most consistent predictor of warning reception is age. The elderly engage in less frequent kinship and community interaction, which is a known factor in limiting the number of warnings older adults receive (Perry 1979). Furthermore, the elderly are less likely to use the internet and social media in comparison to younger populations, and so older adults may be the very last in the community to encounter news of an impending threat. As connections to informal networks decline with age, the elderly may also experience lower rates of reception, as informal networks play a pivotal role in the warning process.

Does the likelihood of receiving a warning then simply decline with age? Ripberger et al. (2019) paints a complicated picture of the issue. When a sample of 3000 respondents were surveyed on attitudes and behaviors regarding weather information, warning reception tended to be lower for younger age groups *and* older age groups, remaining highest among middle-aged groups. This suggests a non-linear relationship between warning reception and age. How might we account for this trend? As noted earlier, for older adults, social circles and networks tend to contract with age, leaving fewer opportunities for interaction, and, by extension, fewer opportunities for encountering warning information through social networks. At the same time, younger people, in spite of having access to a broader range of media, are less likely to watch the news, which means they are less likely to incidentally encounter warning information (Patterson 2007). Age differences also shape the method by which warning information is received. For example, Sherman-Morris (2010), in a study of a tornado warning on a college campus, found that students were much more likely to have received warning information via cell phones while employees were more likely to have received the warning via instant messaging.

In addition to age, Paul et al. (2003) found other demographic and geographic factors that shaped warning reception during quick response research fieldwork following tornado outbreaks among 18 communities in areas across Kansas, Missouri, and Tennessee. Notably, rural communities were far less likely to receive warnings in comparison to urban communities, as were African Americans when compared to Whites. Other factors such as age, gender, and employment status, on the other hand, were found to be statistically unrelated to warning reception. Finally, poverty and low income also serve as obstacles to warning reception. Turner et al. (1979) found that those with higher socio-economic status are more likely to receive warnings.

Finally, a "digital divide," as it has come to be known, makes it more difficult for certain communities to access weather information from computer and digital sources. During Hurricane Sandy, Wang et al. (2019) found that socially vulnerable communities were "digitally left behind" in the context of the hurricane. Other research by Bhavaraju et al. (2019) similarly demonstrated that tweet frequency and circulation declined in some scenarios among more socially vulnerable groups. During Hurricane Harvey, social structural factors also fed into warning response and search and rescue, with research showing that employed, educated homeowners were vastly more likely to use social media to request rescue (Mihunov et al. 2020).

Siren Limitations

Issues also exist with respect to sirens and warning reception. Indeed, sirens are primarily an *outdoor* warning system and are neither intended nor capable of alerting building inhabitants to the threat of severe weather (many tornado sirens are formally labeled "outdoor warning sirens.") Thus, being inside a physical structure during a storm is a major risk factor for not hearing a tornado siren. This is important given that Americans spend nearly 90% of their time indoors, which limits the effectiveness of traditional sirens (Crowe 2010). In spite of these limitations, there is strong anecdotal evidence that the public nevertheless expects to hear sirens while inside their homes (Laidlaw 2010). Thus, an overreliance on tornado sirens to alert communities to the possibility of severe weather may pose significant risks to public safety, especially among those who may depend on sirens as a primary source of warning information. In addition, the geographic distributions of sirens also limits siren effectiveness. Zunkel (2015) found that roughly half of the residents of San Marcos, TX, fell outside the region of siren coverage and therefore were outside the reach of local systems. Time of day is also critical. While only a quarter of all tornado events, nocturnal tornadoes account for 40% of all tornado deaths, a figure likely due to low rates of warning reception (Ashley et al. 2008).

Understanding

Upon receiving a warning message in any form or through any channel of communication, recipients invariably seek to determine what the message means. The process by which warnings are interpreted involves a highly complex psychological exercise because warning messages often express multiple, often-conflicting forms of information, including, but not limited to, information describing the character of the threat, the geographic area over which the threat is present, impact timelines, and threat probabilities. Recipients might ask an endless series of questions: What is the difference between a thunderstorm and a severe thunderstorm? Who issued the warning? What is a tornado "outlook," and how does it differ from a "warning" or "watch?" What is a "short-term forecast" as opposed to a "hazardous weather outlook?" Where am I in relation to the tornado? What does "wind speeds at 70 mph" truly mean? What is meant by "substantial shelter?" Does a "40% chance of flash flooding" mean that flooding is likely or unlikely? As well as countless others. Many may simply be unaware of the common risks within the community, thereby setting the stage for misunderstanding when warnings arise. In a recent study of flood awareness in Colorado, half of respondents were either unaware or unsure whether they resided in a floodplain—an important finding given that those who misunderstood that they were living in floodplain were less likely to respond to flash flood warnings (Morss et al. 2016). More simply, two different people, from two different backgrounds, two different experiences with severe weather, encountering two different warning formats may interpret the risks they face in dramatically different ways.

At what rate are warnings misunderstood or misinterpreted? As it turns out, misunderstanding is not uncommon. Drost et al. (2016: 336) reports that "When asked the typical speed of a tornado and how much advance warning is given before a tornado strikes, only 20%of the respondents answered correctly. Somewhat more disturbing is the fact that only 50% of the respondents understood that a 'warning' is the most urgent National Weather Service Statement." The public is also known to have a difficult time estimating distances relative to predicted storms. In an experiment asking subjects to estimate the locations of tornadic events using historical radar and weather data, Godfrey et al. (2011: n.p.) find that "64% of those who believed that they were not in the direct path of the storm actually were in the path of the storm, with 42% of these individuals unknowingly in the path of a possible tornado." Also reported was considerable confusion regarding the content of warning messages. Subjects often were unsure about the meaning of warning messages: 54% misinterpreted flood warnings and 40% misinterpreted lightning warnings (reassuringly, only 10% appeared to have misinterpreted tornado warnings). Thus, warnings are far from self-evident in meaning and appear subject to a wide range of misunderstandings and misinterpretations.

Past Disaster Experience

A common question asked in prior research relates to whether those with more disaster experience are more likely to understand warning messages. Although Quarantelli (1980a) urges us to approach the relationship between past experience and warning response with some caution, there nevertheless exists a strong and ongoing consensus that those with more prior experience with disasters face less confusion when interpreting warning messages (Guillot et al. 2020; Perry and

Lindell 1986; Foster 1980; Perry and Green 1982; Haas et al. 1977). Recent research appears to confirm this pattern. In one study that examined how college students responded to a recent tornado alert, it was found that those who lived previously in regions prone to tornadoes were more knowledgeable about warnings, watches, and safety protocols (Jauernic and Van Den Broeke 2016). Blanchard-Boehm (1998) links long-term residence, which may be a proxy for greater experience with local disasters, with stronger comprehension of warnings.

Research furthermore suggests that past experience grants better warning comprehension through greater understanding of messages with inconsistent meanings. In this respect, Silver and Andrey (2014) report that prior disaster experience primarily works to resolve conflicts between contradictory forecasts or alternatively clarifies ambiguities perceived in forecast language. Thus, past disaster experience may assist individuals make sense of warnings in two ways. First, by granting a general familiarity with common warning language and concepts—watches, warnings, alerts, etc. For instance, those who live in "tornado alley" may be more likely to understand the distinction between a "tornado watch" versus a "tornado warning" in comparison to those from outside such regions. Second, past experience may also impart the tools to work through warning messages of opposing or differing meanings, allowing those with more experience to determine what is relevant and what is not relevant in the message.

Warning Language, Format, and Content

While the public appears to generally understand that warnings imply risk, the subtle, specialized language in which warnings are communicated elicits confusion (Walters et al. 2020; Silver 2015; Donner et al. 2012; Blanchard-Boehm 1998). In a recent study of wind risks, Williams et al. (2017) found that even among a "weather-salient" sample, there was significant confusion regarding the definitions of a "high-wind watch" vs. a "high-wind advisory." Respondents erroneously reported that the advisory signaled a higher risk than the watch when the opposite holds true. More generally, it was found that the public holds varying degrees of comprehension regarding the definitions of "watches" and "warnings" across a broad spectrum of events (Eastern Research Group 2018: 107).

It is not entirely clear why these terms are so widely misunderstood. It is perhaps due to the fact that "warning" and "watch" are simply too vague to communicate risk accordingly, or perhaps because there is too little overlap between the lay and institutional meanings of either term. In short, although the rate at which the public mischaracterizes watches and warnings remains unknown, evidence at the very least suggests a fair degree of confusion. In the end, what is clear is that not everyone grasps the distinction.

In addition to semantics, confusion may arise due to the number of channels through which a warning is conveyed, as well as the attributes and characteristics of those channels. There is some evidence that the public experiences difficulties understanding the meaning of warnings conveyed solely by way of sirens (Tierney

1987; Lachman et al. 1961). This may be because sirens themselves tend to be vague, or alternatively because the public is simply less likely to understand a warning communicated via a single source, siren or otherwise (Mileti and Darlington 1995). Crowe (2010: 1) draws further attention to and summarizes well the natural ambiguity that attends the use of siren systems:

> Although outdoor warning sirens create a "universal language" at first blush, this language does not necessarily mean the same thing for everyone in a given community. For instance, some local communities activate outdoor warning sirens for high-wind events and some do not (even if they are adjacent communities). This significantly challenges the accepted social science research that indicates that people are most likely to respond correctly if the warning mechanism contains directed protective information and is clear, consistent, and repeated.

Further complexities arise in *how* risk is framed to the warning recipient. The manner in which certainty and uncertainty are phrased is a notable example of this principle. For instance, there is some debate—at present unsettled, though nevertheless important to highlight moving forward—in the literature regarding whether the degree of risk communicated in warnings is more comprehensible to respondents if presented in zero-sum terms (i.e., "the tornado *will* strike") or stochastic probabilistic terms (e.g., "60% chance of tornado")—or as what have been so described as *deterministic* or *probabilistic* forecasts. In some cases, research suggests that probability-based forecasts are well understood by lay people. Roulston and Kaplan (2009) found in an experimental study that groups receiving probabilistic information more often than not correctly predicted 5-day forecast temperatures than those receiving simple, zero-sum information. Notably, the observed patterns remained when controlling for gender and past experiences. Other research, however, shows that probabilistic forecasts of rainfall lead to a greater likelihood of misunderstanding even among a sample of college-educated participants (Joslyn et al. 2009). The matter of deterministic versus probabilistic warning information therefore remains unresolved at the present time with respect to comprehension.

Warnings that evoke stark imagery may aid in understanding, an observation that has fueled discussion over whether individuals find "impact-based" or "phenomena-based" warnings more forthcoming in meaning. Impact-based warnings refer more directly to the likely damage (e.g., downed power lines/trees/building damage) or effects on the person (e.g., traffic delays), whereas phenomena-based warnings discuss severe weather in terms of physical characteristics (e.g., windspeed, scope, etc.). As far as understanding is concerned, Potter et al. (2018) suggests that the public grasps the meaning of impact-based warnings more readily than phenomena-based warnings. This may be because impact-based warnings offer more clarity as to the nature of the threat, whereas phenomena-based warnings are more general and too abstract to serve as a basis for judgement. In other words, understanding may be largely anchored in the ability to visualize a threat in concrete terms. With this said, other authors caution that while impact-based language is generally useful, the use of too dramatic, heavy-handed terms may have an opposite, unintended effect on evaluations of credibility (Morss and Hayden 2010). Thus, if impact-based warnings are used, a focus must be maintained on a language that is at once compelling but not overly sensationalistic.

Finally, demographic factors also shape how the content of warnings is understood and deciphered. Hispanics appear to misinterpret warnings messages and hold less knowledge about local risks more routinely. Aguirre (1988) first documented this pattern in an analysis of a tornado outbreak in Saragosa, TX, wherein deaths and injuries associated with the event were attributed to warnings failing to take into account the "multicultural social context" of the community. Such patterns have regrettably persisted as more recent research seems to confirm the trend that Hispanics hold less knowledge and understanding of extreme weather risks than the general population (Maldonado et al. 2016). Ahlborn and Franc (2012), for instance, in examining the relationship between warning reception and language among Oklahomans, found much lower rates of knowledge of warning systems among native Spanish speaking populations relative to other demographic groups. On the topic of whether minority groups in general are more likely to misunderstand warnings, there is conflicting evidence. Perhaps unexpectedly, Perry (1987) found majority groups more likely to misunderstand a warning message, a finding corroborated by Demerath (1957). However, in Mitchem's (2003) analysis of a tornado event in Indianapolis, it was shown that African-Americans experienced the most difficulty understanding the warning messages issued for the event. On the matter of education, Turner et al. (1979) found schooling to be positively correlated with the likelihood of understanding a warning message, and there is consistent evidence to suggest that older persons are more likely to understand warning messages, as well (Blanchard-Boehm 1998; Turner et al. 1979). If pressed to summarize, it would appear that minorities, particularly Hispanics, the undereducated, and younger people have greater difficult in grasping the full weight of warning messages.

Visualization Formats

Finally, how warnings are presented to the public in visual terms also plays an important role in understanding. In the past, a vast majority of warnings were county based. Today, however, NWS has moved towards the use of "storm-based" or "polygon warnings" These warnings present to the viewer the likely path of a storm, forming a cone of uncertainty, and most importantly take form independently from the county geographic unit (Recall Figs. 2.1 and 2.2 from Chap. 2). Unbound by political geography, polygons may appear within counties, cross county lines, or cover multiple counties. Previously, warnings at the county-level, as implied, placed all individuals and communities within a warned county under severe weather warning— in so doing risking the probability of a higher false alarm rate to reduce the probability of a missed event. Accordingly, a general lack of specificity and geographic vagueness left communities wondering if they were indeed at risk if the threat of tornadoes existed. Furthermore, county-based warnings may have contributed to the "cry-wolf"-effect (in which an overexposure to false alarms breeds skepticism of future warnings), for although individuals may well fall within the perimeter of a county-based warning, they may never actually come to witness the tornado due to the vast size of the area under warning. This could conceivably lead to a belief that the siren, message, or other alert was false even if the unwitnessed tornadoes did indeed touch down somewhere within the county warning area.

The matter of identifying an ideal geographic shape and format for warnings is critical as recent research has highlighted the central role of visualization in the communication of risk (Schumann III et al. 2018). Polygon warnings notably reduce the areas of risk. However, *simply reducing* the size of warning areas from county to polygon may be insufficient to communicate risk to the public because it is unclear whether the public understands the meaning of polygons as intended by the NWS. The NWS states that the likelihood of encountering a tornado is uniform within the polygon's established perimeter (that is, the risk is the same at the centroid as it is for the outer edges). Studies have shown, however, that the public may not fully comprehend this, making errors in judgement about the likelihood of tornadoes occurring within and outside the polygon (Lindell et al. 2016). Accordingly, when polygons were overlayed with radar data, subjects in a later study were found to interpret and respond more appropriately to the storm-based warnings with which they were presented (Jon et al. 2019). Thus, while there is some guidance on polygon-based warnings, additional research is needed before we understand precisely how they are to be used and how they should be presented.

Similar results were found in the context of the visualization of tropical storm warnings. Forecast cones, which are intended to communicate the anticipated path and intensity of hurricane events, nevertheless create some degree of confusion among the viewing public (Broad et al. 2007; Bostrom et al. 2018). In some cases, the cone was interpreted in ways wholly unintended by forecasters. In a passage worth quoting at length, Broad et al. (2007: 663) describe the process by which this occurred:

> The cone of uncertainty is a complicated figure, containing multiple messages represented by multiple graphical elements. Some of these elements, namely, the track line and the cone, have been blamed by some for having contributed (to an unspecified degree) to some public confusion during the 2004 hurricane season. Despite repeated warnings and targeted communication efforts by members of the forecast community, it appears that many people overly focused on the skinny black line (i.e., the track line). Some observed that the line did not pass through their locality and thus incorrectly assumed that they were safe, even if they were still within the cone, indicating that they did not understand that the actual track can vary anywhere within the cone. This suggests that the line actually subverts the key message of a graphic intended to convey uncertainty. Likewise, many people apparently focused on the skinny black line and assumed that only areas along the line were actually at threat, ignoring both the size and severity of the hurricane. The line itself, however, says nothing about how strong or large a storm is. Finally, many people focused on the boundaries of the cone and assumed that if they were outside of the boundaries they would not be impacted. Again, the boundaries only represent a range of uncertainty about the track of the eye. The cone does not provide any direct information about impact risk.

Belief

After an understanding of the warning is achieved (or, at the very least, believing they have understood it), recipients are then presented with a set of questions: will there actually be a thunderstorm in my general area, and, if so, will those thunderstorms likely produce tornadoes, floods, or some other severe weather event? *Is there a risk out there, somewhere, to which I might be* potentially *exposed?* While many questions remain regarding the warning process, and uncertainties abound with respect to why people respond, we know that warnings alone rarely bring about belief in the possibility of severe weather (recall in Chap. 1 the extensive comparison of stop lights, fire alarms, and tornado warnings as evidence of this principle). As will be seen, whether one believes any given warning is often tied to two general factors—the *content* of warning information and *how* that information is conveyed. Regarding *content*, message specificity, for instance, influences whether a threat is believed or not. In the case of *how* warnings are communicated, information conveyed without high confidence or through sources thought untrustworthy may impact the credibility of warning messages. In addition, characteristics such as the legitimacy of the sender, the racial characteristics of the receiver, the efficiency of local warning systems (e.g., false alarm rates) (Atwood and Major 1998; Cola 1996; Breznitz 1984), as well as the environmental and social conditions accompanying the warning, further mold belief in warning information. In general, however, discouraging skepticism tends to be an uphill battle as people bear a tendency to underestimate risks and also tend to regard personal risk as lower in comparison to others. In a series of experiments, Weinstein and Klein (1995) report that individuals are likely to underestimate risks to a wide range of health behaviors due to what the authors refer to an "optimism bias."

Trust in Authorities and Experts

Prior to a warning, the stage is already set for credulity or doubt. Successful risk management of any kind remains heavily dependent on how much or how little faith the public holds in social institutions (Breakwell 2000; Slovic 1993). Indeed, trust in decision makers is known to shape message credibility in a variety of risk domains, including, but not limited to, health risk communication (Pickles et al. 2020), nuclear decision-making risks (Flynn et al. 1993), business-to-consumer transactions (Pennington et al. 2003), and other areas in which perceptions of institutions plays a consistent role in the credibility of information. The origins and character of institutional trust with respect to warnings is neither a simple nor straightforward matter.

While the public likely trusts emergency organizations perceived to hold greater expertise, expertise alone is insufficient to entirely capture public trust. Peters et al. (1997) identify two additional organizational characteristics that appear to shape belief in the credibility of warnings: transparency ("open and honest") and perceived concern and care. Thus, it is not enough that organizations responsible for warnings merely appear technically competent (although that is still an important component); they must also be willing to disclose mistakes and express a genuine regard for the well-being of constituents. This reality is reflected in earlier research on the nuclear industry. In the context of nuclear hazard warnings, the public was found to favor alerts issued from public channels (e.g., government, NWS, Hurricane Center) and remain less likely to trust industrial sources (Flynn et al. 1993). Flynn et al. (1993) documented how the American Nuclear Energy Council (ANEC), an industry group, was not regarded as a credible source of information on nuclear risks in Nevada. This may be because the public interprets industry groups as more likely to downplay risks to preserve industry image, whereas the public sector may be regarded as more transparent and trustworthy in the absence of a clear profit motive. Indeed, the apparent need to trust institutions has been observed by the public's eagerness to build "bonds"—almost quasi-relationships—with television weather broadcasters in whom confidence gradually builds over time to a degree that the "relationship" comes to influence protective action (Sherman-Morris 2005).

Beyond the perceived character of the source, trust in institutions responsible for issuing warnings may be further mediated by individual factors. Siegrist and Cvetkovich (2000) demonstrate that judgements regarding the credibility of warnings is influenced by the background of the recipient. In an experimental study in which subjects were asked to rate risks associated with various activities, as well as trust in institutions, it was found that those with higher perceived knowledge about a risk were more likely to rely on their own judgments rather than trust the judgements of experts. Furthermore, not all institution types are regarded with equal credibility and trust. Local sources of information earn the most trust, followed by state and federal sources respectively, a pattern for which there are two likely explanations. First, local sources and local representatives are, by and large, simply more familiar, and we know that greater trust is placed in familiar sources with which the community has previously bonded or between which trust was previously established (again, see Sherman-Morris 2005). Another likely explanation is that Americans are simply more likely to trust local authorities and comparatively less likely to trust federal authorities on a wide range of issues. Most recently, during the COVID pandemic, for instance, it was found that only 38% trusted federal government with respect to COVID information whereas 55% felt that state and local governments were performing well (https://docs.cdn.yougov.com/pwwefjdnsp/econTabReport.pdf).

Trust in messages may be reduced if authorities are perceived as using language regarded as deliberately evocative or sensationalistic in content or tone. Two days prior to Hurricane Ike in 2008 an official warning message stated that "Persons not heeding evacuation orders…will [may] face certain death." Morss and Hayden (2010) indicate that of the 35 respondents who reported hearing this phrase, roughly 17 regarded the warning in negative or mixed terms, labeling the phrasing variously as "'harsh and over reactive', 'overblown,' 'ridiculous,' 'humorous,' 'stupid, 'rude,' and 'not appropriate'" (186). Casteel (2018) further confirmed that the use of extreme or "scary" language held little effect in the decision-making process.

Thus, while an attempt to provoke an emotional response is necessary to build risk perceptions (which may be why impact-based warnings are so successful), care must be taken so as to not cross the thin, psychological line between belief and incredulity through the use of overblown, unrestrained, or speculative language. Again, the public must *trust* that whoever—be it a person or institution—issues the warning does so in good faith. In other words, the danger presented must not appear contrived. This is important here in another regard. All too often too little or vague information is regarded as detrimental to the warning process. At the same time, caution should be exercised to avoid overly-theatrical language that may negate or undercut the effectiveness of impact-based warnings. If felt manipulated, the public will scrutinize the message.

Environmental Cues

As a general rule, one is less likely to dismiss warnings accompanied by visible severe weather or the effects of severe weather. On the other hand, the absence of any perceptible threats may dilute a sense of risk. Past research suggests the public is much more likely to believe warnings perceptively attended by strong winds, violent thunderstorms, or the very disaster for which the warning was issued (Silver and Andrey 2014; Hammer and Schmidlin 2002; Mileti and Fitzpatrick 1993; Tierney 1987; Sorensen 1982; Quarantelli 1980a; Drabek 1969). Indeed, those within closer proximity to the threat are more likely to believe the message (Sorensen 1982; Diggory 1956), which further supports the notion that people are more likely to believe a message corroborated by visible event conditions. The problem, however, is that by the time environmental cues present themselves, it may already be too late for many to effectively flee, seek shelter, or protect personal property. In the absence of direct evidence of the threat, are there other ways to reproduce the gravity of environmental cues outside of a firsthand encounter?

Consensus is somewhat mixed regarding how to harness the power of environmental cues in the warning format. Some scholarly work suggests impact-based warnings hold the greatest promise in this regard. An emerging body of research in the past several years suggests that impact-based warnings—namely, those in which the consequences of events are communicated or articulated in concrete terms—are more likely to be taken seriously by respondents (Meléndez-Landaverde et al. 2020; Potter et al. 2018; Casteel 2016). Pictures and images are powerful motivators of attention, and their mere presence is often enough to cause the public to take warning messages more seriously (Hu et al. 2019). An argument may be made that impact-based warnings tap into the emotional dimension of human psychology, and, in so doing, provoke feelings of dread that psychologically-flat numerical or probability-based warnings (e.g., "60% chance of a tornado") simply cannot imitate. In this sense, specific formats may more heavily stimulate an effective response. Some scholarship suggests that while audio and visual information did little to trigger an emotional reaction to a hazard, pictures greatly aroused an emotional response among subjects, leading to more acute perceptions of risk (Xie et al. 2011). Pictures may be preferred because it is generally known that verbal descriptions are not well grasped by the public due to the highly contextual nature of human language (Lenhardt et al. 2020; Wintle et al. 2019). Dramatic images suffer much less susceptibility to errors in interpretation and therefore more reliably communicate immanent threats—a picture is worth a thousand words, as it were, when it comes to tornado warnings.

An ongoing debate involves whether communities are more likely to believe probability-based (e.g. "there is a 60% a tornado will occur") warnings or impact-based—("consequence-based") warnings (e.g., "the tornado will overturn cars"). [Recall evidence from an earlier section that probability-based warnings may be better understood by the public. In terms of believing warnings, however, probability-based warnings may be ineffective. Hence, this is why the model is so useful; it calls attention to the fact that similar policies have different outcomes at different stages of the process, again touching upon notions of model dynamics highlighted earlier and to be discussed later.] Ripberger et al. (2015a: 578) cautions an overreliance on probability-based warnings may be the less optimal strategy to warning the public if belief is at stake:

> The logic associated with their design is rather simple—when faced with convective conditions, individuals assess (among other things) the *probability* that a tornado will touch down in their area and the *consequences* that it might have if it were to touch down. If the probability and/or consequences are judged to be high, then individuals will engage in some sort of protective action... Traditionally, NWS warning products focused on the probability side of this equation while neglecting the consequences side. As a result, people were unable to differentiate between high- and low-consequence events, which could lead to suboptimal responses.

But it may be inadvisable to reject probability-based warnings out of hand since there are likely—if presently unevaluated—advantages to both approaches. On the one hand, probability-based warnings might conceivably mitigate the long-term psychological impact of false alarms on the public consciousness since probabilities, by nature, acknowledge the potential for error in judgement (recall in an earlier section that openness and honesty was central to institutional trust, which is essential to promote belief in warnings, and probability-based warnings may appear to communicate more transparently that "this is what we think will happen, but we could be wrong"). As a result, this might be a potential reason that probability-based warnings may be more likely to be believed and responded to (Miran et al. 2020). Yet, on the other hand, a stronger case can be made presently for the use of impact-based warnings due to an overwhelming scientific consensus that environmental cues tend to better corroborate warnings. Impact-based warnings effectively simulate these cues, and more research should be conducted to determined how to best replicate the experience of directly seeing or hearing a tornado, flood, or other severe weather event. Images depicting devastating winds, cars overturned, and people sheltering indeed convey a seriousness unmatched by probability-based warnings.

Message Frequency

It is well established in the literature that people who encounter warnings more frequently are more likely to believe them (Fischer et al. 1995; Sorensen and Mileti 1988; Perry et al. 1982; Baker 1979; Drabek 1969). Indeed, message frequency stands as perhaps one of the most predictable and reliable precursors of credibility and response so documented the scientific literature. Message frequency's impact on belief was a key finding in the seminal NORC study, which was one of the first large scale studies of disaster response (Fritz and Marks 1954). Simply put, those who report receiving more warnings from more sources are more likely to believe that a threat exists. There is little dispute or disagreement in the literature over this assertion.

Yet we do not know exactly *why* the frequency of reception influences belief. To fully grasp and explain the mechanism through which receiving a surplus of warnings increases belief remains difficult given the relationship does not appear to be a straightforward one. The simplest explanation suggests that mere repetition adds a cumulative sense of concern to the point at which the threshold of skepticism is crossed into a state of credulity (Fazio et al. 2019; Deschêne et al. 2010). This straightforward position reflects an argument that belief is a direct, linear function of message frequency. But this explanation may be far too simple.

Although research has consistently shown that the number of sources is positively associated with response, correlation does not reveal causation in the case of severe weather warnings (Luo et al. 2015). For instance, those who are already inclined to believe warnings or who have lower thresholds for risk may simply be more likely to seek out more sources. Likewise, it is possible to speculate that trusted sources—ones more likely to be believed regardless—are simply shared more frequently. What is more, both belief and warning seeking behavior may themselves represent some common cognitive or psychological construct, thereby making the relationship at best spurious. For instance, those naturally fearful of severe weather may find initial warnings warning more credible *and* may incidentally seek additional information, thereby making the observed relationship between warning belief and reception frequency one of mere correlation and not causation. Since so few warning studies apply a longitudinal or true experimental design, or any kind of multilevel approach in the case of cross-sectional analyses, there is great difficulty in factoring out this confounding effect or determining time order.

Method or Channel of Communication

As mentioned, how warnings are conveyed may be just as important as what is conveyed. Specific information is preferred over general information (Lindell and Perry 1987; Carter 1980; Fritz 1957) and warning recipients are more likely to believe sources with which they are more familiar (Mileti and Fitzpatrick 1992), which bear greater consistency (Mileti and Fitzpatrick 1992; Fritz 1957; Clifford

1956), and that are conveyed with greater frequency (Mileti and Fitzpatrick 1992; Mileti 1975; Drabek 1969; Fritz 1961).

In lieu of these insights, there is little consensus regarding what particular source (e.g., the media, emergency management, informal) is most likely to elicit belief. Some evidence suggest that official sources, such as the government and media, enhance belief (Drabek 1994; Li 1991; Perry and Green 1982; Sorensen 1982; Flynn 1979). Other studies indicate that warning belief and response improves when conveyed through personal channels or "significant others" (Li 1991; Perry 1983; Sorensen 1982; Clifford 1956). Social media as a warning tool also has been studied. While social media and crowdsourcing provide easier ways to communicate, people are less likely to trust, and therefore, believe, these sources (Parker et al. 2018).

False Alarms and Warning Skepticism: An Ongoing Debate

Defined as a warning that is issued without a consequent, confirmed event, false alarms continue to generate discussion with respect to impacts on public attitudes towards warnings. At present, there is little consensus over whether, to what degree, and how false alarms serve to weaken public confidence in future forecasts. What can be said, however, is that the influence of false alarms on belief and skepticism is far more complex than is often assumed. According to common logic, it follows that people living within areas with higher false alarm rate will be more suspicious of future warnings, whereas warnings issued in regions with lower false alarm rates face less skepticism. Indeed, there may be at least some truth to this proposition (Simmons and Sutter 2009). Nevertheless, although simple at first glance, the relationship between false alarm rate and skepticism is likely more complicated that might be expected.

We initially might speak of what is known as the "False Alarm Rate (FAR)" in simple, mathematical terms. Accordingly:

FAR = False Alarms/Total Alarms.
With the Success Rate simply being: 1 − False Alarms/Total Alarms.

Thus, if a county issued 100 warnings, out of which 10 were false, the false alarm rate for that county would stand at 10% (10/100 = 0.10). This is the calculable, objective false alarm rate. The argument, which is rooted in two often-cited studies (Atwood and Major 1998; Breznitz 1984), once again suggests that those who reside within high-FAR regions will find future warnings less persuasive. To confirm this, recent research has sought indirect evidence in the form of impacts of false alarms on mortality. In perhaps the most methodologically-sophisticated study of the impact of FAR on tornadic casualties, Simmons and Sutter (2009) found that a one standard deviation increase in the ratio of false alarms to true (confirmed) alarms may be expected to increase the likelihood of fatalities from 12 to 29% even after controlling for factors related to lead time, social conditions, and storm intensity. The false alarm effect is further seen in Trainor et al. (2015), the results of which confirm that objectively-measured false alarm rates do indeed correlate with observed behavioral response to tornado warnings.

More recent scholarship, however, raises more questions than it answers about the relationship between belief and severe weather false alarms. In a survey of residents in the southeastern United States, Lim et al. (2019) demonstrate variation between public subjective judgements of false alarm rates and the actual false alarm rates of counties. Counterintuitively, results reveal that respondents who perceived the false alarm rate as being lower were more likely to take protective action in some form, leading the authors to conclude that "Residents' perceptions of false alarm ratios and tornado alert accuracy appear to matter more than the actual false alarm ratios and the number of tornado warnings they have received when it comes to their behavioral responses." (560). [Although note in Trainor et al. (2015) the effect of perceived false alarm rate was far less pronounced].

Additional research by Ripberger et al. (2015b) adds additional uncertainty. According to the study, it is not merely the objective false alarm rate, but also the frequency of missed events [i.e., the Missed Event Rate (MER)] that shape an overall sense of trust in warnings, which, in turn, was observed to shape the intended behavior of respondents. Overall, it was found that while high false alarm rates and high missed event rates shape "subjective perceptions" of FAR and MER, social and interpersonal factors also independently explain perceptions of false alarm rates. For instance, age, education, race, and gender shaped variously the perceived False Alarm Rate and the perceived Missed Event Rate above and beyond what these rates were in reality. Thus, while the overall false alarm rate and missed events guide perceptions of the false alarm rates, such perceptions are further molded by outside factors. This would seem to suggest that while the objective false alarm rate and missed event rates shape perceptions, perceptions are not ultimately reducible to these rates.

Demographic Characteristics

Finally, as with pervious stages in the warning model, demographics play an important role in shaping perceptions of warning credibility. Women are more likely than men to believe warnings (Schumann III et al. 2018; Farley et al. 1993; Drabek 1969), and this is perhaps not surprising given that we know that women are, on average, less likely to take risks relative to men across different domains of risk taking (Croson and Gneezy 2009; Harris et al. 2006; Byrnes et al. 1999), among which is included warning response. Another reason may be related to the fact that persons to whom children are entrusted are more likely to respond to warnings (Perry 1981), and women still today are overwhelmingly more likely to be primary childcare providers. Likewise, older adults are less likely to believe a warning (Hodge et al. 1981; Mack and Baker 1961). While there is no strong explanation for this pattern, it is possible to infer that older adults, because of potential overexposure to false alarms throughout the life course, and all else being equal, may over time grow more wary of warnings in comparison to younger populations. Finally, persons of higher socio-economic status are more likely to believe warning messages (Perry 1987; Sorensen 1982) and membership in a minority group reduces the likelihood of believing the message (Nigg 1987; Turner 1976). One should ask, of course, whether these latter findings point to the role of education in believing a warning, as minorities and the poor tend to have fewer years of schooling. Interestingly, a recent survey found that Hispanics were more likely to shelter when warned by a siren, but less likely to trust and respond to English-language warnings from the same authorities (Stokoe 2016).

Threat Confirmation

At this stage in the model, the warning recipient believes that the message is indeed credible and there is cause for concern, but the risks, although now regarded as *somewhere reasonably close*, remain too vague to warrant action. Action rarely results at this point because the costs of acting immediately upon a mere general belief are often judged too great—*and yet*, the risk cannot simply be ignored without a further investigation and appraisal. Confirmation serves as yet another means by which to avoid anticipated costs of response, sheltering, or evacuation (Whittaker et al. 2020). Warning recipients seek to resolve this tension through confirming warning information and/or by confirming the immediate presence of severe weather (the latter of which leads to personalization, the next stage in the model). In the literature, confirmation is a process that rarely goes undocumented and may take place through a variety of means: revisiting formal warning information, trading information with friends and neighbors, or simply looking outside. What must be understood is that people do not live in a "risk bubble" where warnings take precedence over other social obligations and risks, but, rather, warnings form part of a set of risks and obligations out of which only a finite set can be acted upon at any given time. Do I respond to the warning by staying in my home, or ignore the warning to fulfill other obligations: go to work, drop my children off at school, or attend an important social event? Naturally, these activities cannot be done simultaneously and so must be ranked in terms of importance in the minds of warning recipients. There is no guarantee that the warning information will be given precedence. And so initiates the process of confirmation to determine if it should.

It is necessary to ask, therefore, what does confirmation *do?* Which is to ask, what is the *function* of confirmation in the warning response process? The act of confirmation seeks to minimize ambiguity in a variety of areas. Is the information true? Are reports consistent or do they conflict? Where is the tornado likely to strike? Can the tornado be visually confirmed? Can others be seen preparing or responding? Social pressure may give way to the need to confirm warning information. For instance, Griffin et al. (2004) report that individuals seek out additional risk information not merely due to their own concerns, but also because they believe others would expect them to. Indeed, more recent research confirms this. Gutteling et al. (2018) show that social pressures resulted in a greater likelihood to respond, confirm, and monitor the environment when threats were encountered. Here it is important to note that although initial perceptions of credibility (in messages, institutions, etc.) may play a role in shaping belief about the truth of warnings (Perry 1987), warning confirmation often relies on interaction with informal groups as a mechanism by which personal threat is assessed. Whereas people are more likely to believe *official sources* about evolving natural hazards (as with the previous stage), the transition to risk personalization is preceded by confirmation (Perry and Green 1982) achieved through a synthesis of both formal and informal sources, including authorities, friends and family, and environmental cues (Parker et al. 2007; Drabek 1969, 2000). As it were, the process of confirmation is far more "messy" than the initial evaluation of truth described in the earlier stage.

What are some common patterns in the reception process? Research suggests that the more times people receive warning information, the more likely they are to confirm it (Nigg 1982), a conclusion for which two explanations may possibly exist. One explanation for this observation might be that *ceteris paribus* encountering multiple warnings increases the likelihood of receiving conflicting information, which, in turn, leads to a subsequent search for additional information to resolve the resulting ambiguity. Interestingly, and perhaps somewhat counterintuitively, the more specific a warning message is, the more likely the receiver is to confirm it (Cola 1996). Stokoe (2016) reports confirmation to be somewhat common when sirens are encountered. When asked about how they normally respond to tornado warnings, 40% of respondents report immediate action. Nevertheless, 30% of siren recipients reported seeking out other sources of information for confirmation and 25% reported looking out the window to confirm. In the same study, confirmation reveals itself to be demographically complex:

> Hispanics were significantly more likely to seek immediate shelter when they heard sirens (47%) while those over 60 were more likely to check other sources of information (39%). This variation reversed when they received WEA alerts, with Hispanics more likely to check additional sources (70%) and older people less likely to shelter (33%). Overall, women were more likely to take shelter than other groups, while white men were consistently more prone to ignoring the alerts, looking from a window, going outside, or tornado chasing (146).

Personalization of Risk

The *function* of confirmation is to act as a sort of bridge between general belief and personalization in the stage model of warning response. In many cases, warning messages fall short of eliciting response simply because the risk they convey is too uncertain and abstract—the tornado, flood, or severe thunderstorm will likely happen, *but likely happen to someone else or somewhere else*. In other words, to begin setting the stage for protective action, recipients must *define* the situation as one that is immediately dangerous (Ponting 1974). Thus, merely acknowledging the presence of a general risk is not tantamount to believing that a risk is immediate and immanent. Personalization of risk implies the assessment of whether one is in danger of experiencing a disaster event and the degree to which one will be affected. What factors shape the personalization process?

Cultural Myths

Just as we inherit many of our beliefs about religion, politics, and social norms, so too do we inherit our beliefs about disasters from generation to generation through the mechanism of community socialization. No mind is a blank slate, as it were, as far as the weather is concerned, and perceptions as to the origins and character of storms shape perceptions of immediate threat. While previous anecdotal research suggests a possible role for cultural myths in the disaster warning process (Donner 2007), this view has recently been corroborated in the quantitative literature (Schumann III et al. 2018). But, precisely, how do local cultures influence the disaster warning process? Notably, cultural myths appear to have some impact on personalization. In some cases, local cultures distort understanding of natural phenomena. For example, some communities regard certain features of the local geography to offer absolute protection from severe weather hazards. Mountainous or hilly regions, commonly thought to shield communities from tornadoes by acting as natural barriers, nevertheless experience a fair share of severe weather (Klockow et al. 2014). Indeed, such myths are far from rare. One recent study reports that 90% of respondents regard at least one feature of geographic significance as a safeguard against severe weather (Jauernic and Van Den Broeke 2016). Thus, a person ascribing to such sentiments may believe a warning message but may not feel personally threatened. Stated succinctly, warnings "must be perceived by [communities] as relevant to their situation", and believing that geographic features offer protection may indeed make warnings irrelevant (Rodríguez et al. 2006: 480).

Cultural Fatalism

An often overlooked feature of warning response, and one deeply relevant to the personalization process, are feelings of control (or the lack thereof) over the outcomes of severe weather. While "cultures of resilience" are not uncommon among some communities (Aguirre 2004), there are nevertheless circumstances under which "threatening situations can be incorporated into the familiar schemes of 'normal' cultural problems [whereby] [t]he nature and degree of accommodating the schemes then becomes critical in analyzing the cultural adaptation to threatening situations and the consequences for structured behavior in crises" (Anderson 1968: 303). To paraphrase, risks cease to evoke concern because the risk in question becomes "normalized" or inevitably connected to day-to-day activities. Often because the risk cannot be mitigated or responded to in an effective manner (either because it is inevitable or merely perceived so), the community eventually arrives at the conclusion that any course of action intended to build resilience will ultimately be futile—a passive attitude comes to prevail.

In colloquial terms, this phenomenon is often labeled "fatalism," and has been shown to play into how communities respond to potential threats. Nevertheless, the roots of a fatalistic worldview with respect to disaster are not well known. Some research suggests that the frequency with which disasters occur teaches victims to accept the inevitability of disaster. For instance, Fritz's (1961) seminal writing on disasters and society offers a lucid description of how mining communities often come to accept as inevitable mine collapses due to their high rate of occurrence. Fatalism may also be correlated with different subcultures or groups within society. For instance, there is evidence to suggest that less educated groups are more fatalistic and pessimistic than their better-educated peers (Maercker et al. 2019). Under such circumstances, while there is some belief that risk may be present, culture undercuts risk personalization if communities believe themselves helpless in the face of threats.

Social and Environmental Cues

Earlier it was noted that environmental cues increased the likelihood that warnings would be believed. The literature suggests the same may hold true for the personalization stage of the model. There is also ample evidence that social interaction ("social cues") at the time of the warning increases the tendency to personalize risk. Once one believes a risk is present, cues, either originating in the natural or social environment, must be witnessed if such a belief is to be personalized.

Repeated studies of earthquake response demonstrate that the likelihood of preparing for a disaster increases when others are seen as doing so (Mileti and Darlington 1997; Mileti and Fitzpatrick 1992), and it has been shown that the likelihood of evacuation is greater among those who witness others evacuating and business closure (Lindell et al. 2005). The authors speculated that this relationship may have been simply due to the greater likelihood of warnings in visibly impacted areas, but, after controlling for the presence of warnings and demographics, the effect of social cues remained significant. This would seem to suggest that social cues have an independent effect on personalization across individuals and may be linked to environmental and social cues in the immediate environment. Mileti and Darlington (1997: 98) have ruled that social cues function to "convert new information into action." Seeing neighbors evacuate, witnessing business closures, and observing the presence of spotters all seem to have a profound impact on the ongoing sense of normality to which the community often clings even in the face of multiple official warnings.

Finally, cues may offer one explanation as to why geographical proximity to the event is so often found to lead to the personalization of warnings (Sherman-Morris et al. 2020; Flynn 1979; Perry and Lindell 1986; Rogers and Nehnevajsa 1987) It follows that if one is within closer proximity to the hazard and in a position to witness their effects, then one is more likely to directly experience those cues through which personalization is elicited.

Warning Content, Past Experience, and Demographics

Finally, personalization may stem from a number of other causes. For example, elements of the content may impact personalization, although the direct effects on personalization is not well understood. Personalization of risk appears to follow from warning consistency (Lindell and Perry 1983; McDavid and Marai 1968), as well as the specificity with respect to the message (Perry et al. 1980). Not surprisingly, a sense of personal risk is likely to originate from sources viewed as more credible (Perry 1979; Rogers and Nehnevajsa 1984). Despite considerable disagreement among scholars (Mileti and Darlington 1997), there is general evidence that past hazard experience makes it more likely for one to derive a sense of personal risk from warnings (Hansson et al. 1982; Rogers and Nehnevajsa 1987; Saarinen et al. 1984; Perry 1979). As with prior stages of the warning response subsystem, personalization of the risk process varies widely with demographics. Older adults are less likely to personalize warning messages. Reflecting prior trends with respect to gender and race, women and whites are more likely to personalize risk (Hodge et al. 1981; Mileti and Darlington 1997).

Believe Protection Action Necessary

Following personalization, and now in fear of an imminent threat, those warned ask themselves, "Must I do something to make myself safer or am I presently safe enough?" Unfortunately, there is minimal guidance from the literature about how one evaluates the necessity for protective action. This difficulty stems from a regrettable scarcity of empirical research.

Before we proceed, there is a need to distinguish the present stage from earlier stages related to belief and personalization. Prior stages dealt with the credibility of warnings and estimates of personal harm, whereas this stage addresses psychological processes governing the assessment of whether the current situation demands action. In some cases, one may feel at risk yet believe that additional action would do little to enhance one's safety. For instance, one may believe that the authorities or some organization can appropriately manage the threat, that the geographic region in which one is residing negates the threat, or that taking action would actually increase personal risk. In simpler terms, *one now feels unsafe*, but the question now is whether action is necessary to make one more safe.

Some may decide further action unnecessary because the outcomes of the disaster are regarded to be out of one's control. In radical cases, victims come to view deities as protective against disasters (Moore 1958), such that any behavior of the individual will be viewed as inconsequential within the context of a larger set of deterministic forces. Other cultural factors may play a role. For example, in an analysis of the flooding of the Rio Grande, Anderson (1968) found cultural prioritization of honor and self-sufficiency—two factors that may either reflect or foster a greater sense of personal control—among Mexican communities as obstacles to seeking assistance from Anglo communities. More troubling are those cultures in which protective action is not only deemed unnecessary, but also undesirable. Anderson reports that the survivors of New Guinea's eruption of Mt. Lamington "expressed feelings of guilt and insecurity because they interpreted the event as punishment from God" (301).

Perceived locus of control—the degree one believes one holds the efficacy and agency to bring about change--appears to factor heavily into the evaluation of whether protective action is necessary. The relationship between locus control and the necessity of action in response to disaster has until recently been largely grounded in speculation and anecdote. However, more recent research demonstrates strong links between self-efficacy and disaster response. In a study of volcanic eruptions in Mexico, López-Vázquez and Marván (2012) found that respondents who ranked high on a "fatalism/luck" scale (taken to imply low locus of control) tended to adopt more passive approaches to disaster preparedness, suggesting an aggressive need for preparedness to be lacking. Baumann and Sims (1978) earlier found that locus of control played a significant role in the decision to adopt flood insurance, as did Spittal et al. (2008) similarly, but in the context of earthquakes. Here the point is that one can *clearly* believe that one is at risk, but, because one lives in a culture in which life, death, and injury are regarded as largely pre-ordained or random in character, one may feel little efficacy or compulsion to act. *The lesson here is that mere feelings of personal risk do not* necessarily *give forth to a belief that a course of action will, in the final analysis, reduce risk.*

Trust also shapes belief in the necessity to act on warning information, but in unexpected ways. Note that whereas *deficits* of trust in institutions held negative outcomes for warning belief, too much trust in institutions may hold negative outcomes when evaluating whether action is necessary. Illuminating this point, in a series of in-depth interviews in Korea regarding flood risks, Kim (2019) discovered that perceptions of institutional competence among communities may have a cooling effect on response. According to the findings, many of the respondents interviewed did not take protective action because they believed that authorities could manage the problem effectively. Thus, while too *little* confidence in institutions may lead to inaction, the opposite may also hold true: too much confidence in institutions may lead to feelings of security where none exists, which, in turn, creates a false sense of wellbeing, and inaction. This example illustrates well the idea that even if people *believe* a risk is present, and feel personally at risk, excessive faith in institutions may lead to diminished urgency to act. Earlier research further suggests that excessive confidence in authorities leaves the community unprepared for unexpected events (Janda et al. 1996).

Finally, it should also be noted that caregivers are more likely to take action once risk is established. For example, households with children are more likely to evacuate from and prepare for hurricanes (Dash 2002; Gladwin and Peacock 1997). This would seem to suggest that altruistic behavior such as caregiving responsibility might motivate response, and therefore may extend to disability and elder care, as well.

Believe Protective Action is Feasible

A related question is whether protective action is ultimately feasible. A recipient may feel the *need* to take action, but, at the same time, such action may be regarded as impossible, unwise, or unsound in final judgement. Many who intended to evacuate from the Gulf Coast prior to Hurricane Katrina—an action they deemed as necessary—failed to do so as evacuation was thought to be impractical (and indeed was for many of its very poor residents). There is also an element of perception involved in evaluations of feasibility; resources for protection may be readily available, but those responding may be unaware of them or unaware of their utility during a disaster. For instance, a house may have a bathroom, but the resident may be unaware of its utility as a tornado shelter. The local elementary school may convert to a shelter during hurricanes, but not everyone may not be aware of its availability. Others may choose not to evacuate as they fear their homes might be looted. For instance, during Hurricane Gilbert Aguirre (1991) discovered that fear of looting largely explained people's reluctance to respond to evacuation orders. Naturally, if one fears that their home will be stripped of possessions once abandoned, then one might view evacuation as infeasible or unwise *even if* the very act of evacuation was regarded as necessary. Many may elect to neither shelter nor evacuate if the costs of doing so—potential theft, vandalism, and destruction of property, loss of time, loss of work income—exceed the anticipated returns on safety.

Evaluations of feasibility are often grounded in judgements related to the *costs* of evacuating from a hazard. There are many risks one faces on a daily basis—not just potential severe weather risks. Put another way, those warned balance the risk of breaching social and economic expectations, on the one hand, against the risk of death or injury from the event for which the warning was issued (Donner 2007). There is presently no law or punishment for refusing to heed a warning. One cannot be given a jail term, cited, or fined for ignoring sirens or NWS warnings, nor are businesses required by law to shutter. Therefore, the burden of weighing the benefits of response against the costs of doing so falls on the shoulders of the recipient. To that point, Drabek (2001) reported that among employees a significant factor in warning response was the conflict between professional and personal roles. Does one risk missing work to respond to a warning? Does one risk missing important classes to respond to a warning? Or a major and important social event such as a party or wedding? These may seem like minor or trifling concerns, but in the end they nevertheless compete with the decision to act upon warnings. Disregarding social or economic obligations, such as work and school, may come with costs that outweigh the act of sheltering from a storm that may or may not occur. In a study of residents following Three Mile Island, Houts et al. (1984) found that among those who did not evacuate, nearly a quarter (22%) identified some social or physical barrier to response, including being unable to leave work, lack of transportation, lack of place to go after evacuation, and physical disabilities. The draw of other important activities, events, duties, and commitments often eclipse the urgency tornado warnings create, as well. Evaluations of feasibility, therefore, are rarely made in a social or psychological vacuum.

Protective Action: Resources and Lead Time

Often oversimplified as "evacuation" or "shelter seeking," protective action is not a single activity, but rather a series of value-added activities oriented towards the reduction of risk. The process incorporates both the real and perceived presence of resources. In these final stages of the model, warning recipients feel not only an acute sense of personal risk, but also believe that action is necessary and feasible to mitigate that risk. Two conditions must exist to enter this present stage of the model: (a) *those warned have personalized the risk* and (b) *arrived at the conclusion that doing* something *offers a greater chance of survival than doing nothing*. Thus, recipients have selected what they believe is the optimal course of action—which may be no action at all. Yet what is regarded as necessary and feasible may not be viable. One can define a situation as unsafe, but cannot take action if resources are insufficient or unavailable. In order to transition from a psychological state to taking protective action, in whatever form, resources must be available and be known to the warning recipient. We know that resources are rarely available to all members of the community equally.

On the Availability and Use of Resources to Assist in Protective Action

Patterns of resource availability and use is a complicated problem that operates across multiple levels of the social order. Resources that aid and factors that affect protective action differ in distribution at the *individual* level (e.g., by income, age, disability), *community* level (e.g., shelter or sirens, trained emergency managers, media coverage in rural vs. urban areas), and *geographic* level (coastal erosion, population density, rarity of the hazard). Furthermore, while resources may commonly be considered in terms of public shelters, trained emergency managers, a vehicle to support escape, or money to sustain long-term evacuation, here we regard knowledge as a resource of equal status to physical resources as far as facilitating protective action is concerned. All resources—from the tangible to the intangible—form what might be regarded as "capital" (Meyer 2018) in the response process, and capital may be regarded in a variety of ways, from money, through social network capital, through cultural knowledge of what to do during a severe weather event. Indeed, all protective action takes place within a specific context and is aided through a combination these resources after one decides it is time to take action.

At the *community* level, race, class, education, age, socioeconomic status, geographic location, and frequency of the hazard among other variables, shape the availability of resources useful for response, preparedness, and mitigation [see Donner and Rodríguez (2008) as well as Fothergill and Peek (2004) for a somewhat dated yet nevertheless comprehensive overview of literature on these topics]. Education and income have been tied to resource availability during storms in areas where severe weather is common. For instance, Balluz et al. (2000) find that those with at least a high school education and persons with basements were more likely to have access to shelters. Edwards (1993) reports similar findings in her study linking higher education and higher household income to the likelihood of having made earthquake preparations. Cultural norms, which originate at the level of community and govern the behavior of members, operate in much more subtle ways and may also give rise to differing "styles" of response. In their examination of the structural and cultural differences between volcano victims in the United States and Japan, Perry and Hirose (1991: 112) find that the Japanese were more likely to seek public shelter. The researchers hypothesize that this may have been due to the difficulty of sheltering large families in Japanese households (which are traditionally small) and the existence in Japan of a "collectivist culture in which citizens have higher expectations that authorities will provide care in the event of disasters or other disruptions in social life."

In conceptualizing resources at the community level with respect to protective action, we should also consider routes of escape as well as shelter quality. Are escape routes generally known, maintained well enough to support evacuation during severe storms, and easily-accessible? Or are most escape routes congested and difficult to access? Has a power outage during the storm disrupted communications between emergency management, the media, and the public? Furthermore, the degree to which shelters provide comfort has been found to play a role in the decision to shelter. For example, Cola (1996) finds uncomfortable shelters less likely to be used during storms, and many local communities do not accommodate pets (Farmer et al. 2016), which may deter shelter seeking for some. Naturally, if a high-impact event is nearby, recipients observe clear indications of its presence, and every source confirms its impending impact, shelter comfort may have much less influence on the decision to evacuate. But if there are high levels of uncertainty, or if there is a high false alarm rate within the community, then uncomfortable or unaccommodating shelters may deter some from considering these facilities. Again, as was seen, protective action always involves a weighing of risks and never takes place in isolation.

Finally, *individual* factors come into play. The mere existence of resources at the community levels does not imply nearby recipients will know of their existence. Those most likely to be unaware of resources could conceivably include recent residents and visitors (including tourists) to an area. As with the previous stage, financial, social, and cultural capital impose limits on the knowledge of resources one is likely to possess. New residents and visitors lack the specific a familiarity with disaster that defines the lives of many local residents. For example, new immigrants may lack an awareness and understanding of regional hazards, the protective resources available to the community, and the social networks through which both could be obtained. Those most removed from local disaster culture are at significant risk.

Lindell and Perry (2004: 58) suggest that "perceived implementation barriers that affect protective action decisions arise from resource constraints that preclude the selection of a preferred protective action, as well as obstacles that are expected to arise between the decision to take a protective action and the achievement of protection." Knowledge of safe areas and how to go about accessing them may constitute a major barrier to effective protective behavior (Perry et al. 1980; Windham et al. 1977). Again, it is those who lack strong cultural capital and embeddedness within social networks who appear to be most lacking in knowledge and understanding of these resources. While the location of shelters or other places of refuge may simply be matters of common sense to local residents, new residents, tourists, and other outsider groups may hold scant knowledge of their existence. Moreover, members of minority cultural or subcultural groups, lacking communication with emergency management institutions, or formal institutions more generally, may not know the proper procedures or resources by which to protect themselves. For instance, new immigrants and tourists may be unaware of the location of local shelters.

Finally, there remains the physical act of response, the successful completion of which may be prevented by a wide range of factors. For example, persons with developmental disabilities require special training and needs in cases of fire evacuation (Ramey-Smith and Fletcher 1978). Pearson and Joost (1983) later discovered through the analysis of several fire scenarios that carrying out activities associated with evacuation, e.g., calling a fire department, searching for a young child (i.e. a small doll), and donning a robe, took blind participants 2.47 times longer and wheelchair participants 2.36 times longer than non-disabled participants. Regarding age, in the same study older adults took 27% longer and those with arthritis took 44% longer to evacuate in comparison to younger, healthier subjects. Other studies demonstrate that the disabled and elderly, due to mobility issues, remain at a distinct disadvantage during disasters. In a study of 65 nursing homes, Voght (1991) found several key factors that influenced the amount of time required for evacuations. These included the type of threat, the availability of sources to help, and population density of the county in which the nursing home was housed. Surprisingly, the percentage of non-ambulatory clients, the presence of a disaster plan, and the average age of clientele were found to have little impact on the evacuation time. These results suggest that institutional assistance may serve to partially reverse the negative impacts of age and disease that might otherwise prevent individual evacuees from egressing in a timely manner (as was seen with the previous two studies). However, for large scale mass evacuation, such as flood events that require a sustained, physically-demanding passage through high water (Lee et al. 2019), it should be noted that these findings may not hold (Nomura et al. 2016). Where the surrounding environment places great demands on the physical and mental limitations of elderly, institutional assistance may be an absolute necessity. Thus, isolated elderly populations who live outside an institutional setting would appear to be least likely to take protective action and therefore suffer high levels of vulnerability during a severe weather event. Aside from age, Hoekstra et al. (2011) report that those with no children were far less proactive when it came to acting upon warnings, being less likely to evacuate, less likely to confirm warning information, and less likely to contact and coordinate with family members.

Another relevant factor that may limit the final act of response but which at the same time is often not considered is pet ownership. Some public shelters prohibit animals, leaving the pet owner with the difficult decision of leaving his or her pet behind or risk remaining home during a threatening event. Although it would appear that no reasonable person would refuse free access to shelter on account of a pet, in fact it is not uncommon for people to forgo evacuation in the absence of pet accommodations (Pfister (2002). Among non-evacuees during hurricane Katrina nearly half reported not wanting to abandon pets as a significant reason for not making use of shelters (Fritz Institute 2006). This matter became so important for shelters following Hurricane Katrina that FEMA and the military had to change their policy to accommodate pets in the evacuation process.

Conjectures and Propositions Regarding the Role of Leadtime in Protective Action

In closing the discussion on the warning model, it is necessary to discuss lead time, formally defined as *the time between when a formal warning is issued and the time at which the event for which the warning was issued strikes the community*. For example, under this definition, if a community receives a warning at 10 am, and the tornado strikes the community at 10:10 am, then it is said that the community in question had 10 min of lead time to respond. Lead time factors heavily into discussions of severe weather safety under the basic assumption that the longer one has to hear, understand, and physically respond to a warning, the safer one will be. The simple logic is that communities will be safer if lead times are longer and less safe under shorter lead times. In reality, however, the relationship between lead time and human behavior remains a far is a more complex affair. Apart from a few large-scale demographic studies [see Simmons and Sutter (2008) and Donner (2007) as examples], we know very little about actual behavior during varying lead time conditions. This may be for two reasons. First, there are fundamental differences between official, community, and individual lead times. Second, the relationship between lead time and safety may not be linear but rather curvilinear.

Official vs. Community vs. Individual Lead Times

As far as warning response is concerned, and contrary to the original definition, there is no *single* lead time, but rather several interrelated lead times: *official lead times, community lead times, and individual lead times*. Accordingly:

1. Official lead time: The time between the official warning (e.g., siren, NWS warning) is *issued and* the time of hazard onset.

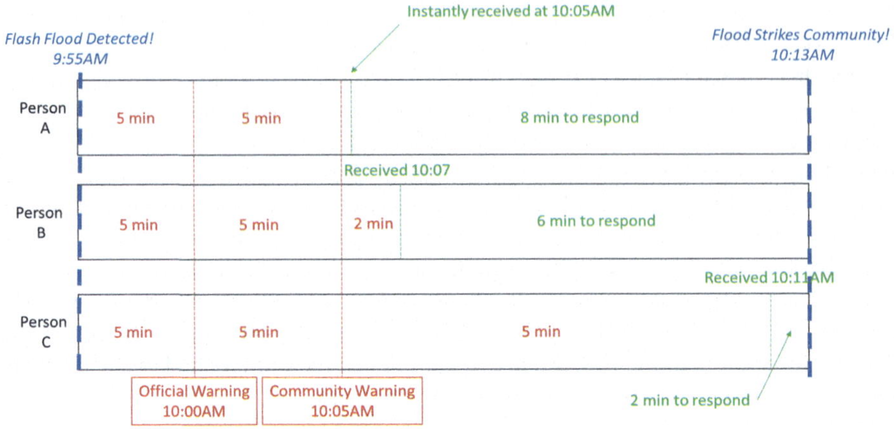

Fig. 3.2 Differences between official, community, and individual lead times of individuals in same community

2. Community lead time: The time between when the community is *aware* of the hazard and the time at which the hazard strikes the community.
3. Individual lead time: The time between when the individual is *aware* of the hazard and the time at which the hazard strikes the community.

All three may happen simultaneously, but, more often than not, there is a lag between the official warning, community awareness, and individual awareness.

For instance, assume the following scenario (See Fig. 3.2). The NWS may issue a warning at 10:00 am for a flash flood. Local news begins broadcasting the.

warnings at 10:05 am. Already watching the news, Persona A therefore immediately receives the warning at 10:05, whereas Person B only sees this information at 10:07 am after sitting down to watch television, and Person C at 10:11 am after a friend calls to alert him or her to the warning. The flash flood or tornado eventually strikes the community at 10:13 am. Under the standard, operational definition of lead time, each person would have 8 min of lead time from the point at which *the community* received the warning. However, not everyone received the warning at the same time under this scenario Accordingly, Person A had roughly 8 min of lead time; Person B, 6 min; and Person C, 2 min (see Fig. 3.2 for an illustration). This is highlighted because the warning model we apply in this volume initiates—that is, from *reception* onward—not from the point at which the warning is officially declared or issued to the community, but rather when an individual member of the community hears or sees the alert either directly or through social networks. Thus, in terms of an interdisciplinary model, factors that affect individual reception are key.

An Unanswered Question: What Is the Ideal Lead Time?

A solution to the problem of "lag" between the official warning and individual response might be to simply extend the initial warning lead time through the use of technology to an indefinite period—hours, perhaps, to ensure that everyone has enough time to prepare, shelter, and evacuate. The relevant question here is: do longer lead times lead to more effective protective action and ultimately fewer deaths and injuries? One assumes that *the longer the lead time, the safer one will be* (i.e., the probability of death or injury from a tornado event is inversely related to lead time length) and vice versa. If this held true, we would generally see an indirect, linear relationship between lead time and mortality or morbidity. For this to be true, warning response must be governed by a "stimulus-response" model of psychology in which individuals immediately and unequivocally recognize threats implied in warnings and accordingly take protective action for as long as the warning remains in effect. If such assumptions were correct (and scholarship suggests they are likely not), then those warned would immediately seek and remain sheltered for the duration of the event, emerging only after the hazard was no longer a clear and present threat. Unfortunately, we know little *directly* about behavior under varying periods of lead time. This is in no small part due to the difficulty in measuring lead time in public surveys (studies using secondary data have been more successful though tend to be limited to the geographic levels of analysis, but tell us very little about the consequences of varying lead times for the individual). The main source of difficulty arises because survey respondents would be hard pressed to recall with any measure of reliability or validity the period of time between when a warning was received and when the tornado struck. Memory limitations, the role of stress in distorting recollections of events, and limitations in knowledge of when and where the tornado struck limit what we can reliably know about personal lead time. Thus, little research has taken place on the matter of lead time at the individual level of analysis.

There is indirect evidence, however, suggesting that there may be a general unwillingness to tolerate lead times that are regarded as too long because people may be disinclined to shelter for extended or indefinite periods. In an attempt to evaluate and provide data for NOAA's experimental "warn-on-forecast" approach, Hoekstra et al. (2011) evaluated attitudes towards lead times among a sample of 320 members of the community. On average, respondents said that around 34 min of lead time would be ideal, although the authors report that there was some difference in preferred lead times by region. Roughly 40% of respondents were noted as claiming that there could be too much lead time, a figuring mimicking earlier work in which roughly 37% of respondents stated similar views (Ewald and Guyer 2002). Hoekstra et al. (2011: 137) attribute the belief that there could be too much lead time to the fact that "A possible explanation for why a longer lead time could be too much might be because the individual feels the added time is inconveniencing the rhythm of their day. This extra time would not be as valuable if the individual waits until the last minute to seek shelter anyway." Other research cautions that excessive lead times may, contrary to intentions, encourage risky behavior. For instance, if given sufficient lead time, some individuals may attempt to flee the storm as they have been shown to do in some cases (Hammer and Schmidlin 2002) and as a result may be killed or injured after encountering the storm unexpectedly (for tornadoes are one of the most unpredictable forms of severe weather). Again, although there is little research on the matter, what little research does exist suggests the need to rethink the basic model of lead time under which present policy operates.

The Question of Model Dynamics

Mileti (1999), Mileti and Sorensen (1990), and Lindell and Perry (1992) offer a model much more faithful to the realities of how people respond to risk in comparison to simple, one-dimensional stimulus-response explanations. While the model holds great descriptive utility, little is known about the process by which stages transition. That is, scholars know far more about what happens *within* each stage and less about what happens dynamically *between* stages.

Hypotheses About Stage Transition

The warning response model appears to assume that stages transition in a linear manner—from reception to protective action in a linear and continuous fashion. But there are grounds for hypothesizing that may not be the case, for the behavioral and psychological realities that drive warning response are deeply complex and contextual. To address this complexity, three hypotheses about the dynamic nature of warning response are offered and briefly discussed:

1. *Dependence*: stages do not progress independently; that is, decisions made in earlier stages impact later stages;
2. *Non-linearity*: stages do not transpire in a unidirectional manner, but are recursive and often reflect decisions made at later stages; and
3. *Differential Variable Impact*: social variables, which affect the outcomes of each stage, do not impact each stage in the same way or to the same degree.

Dependence

Evidence implies strongly that the events of earlier stages shape outcomes in later stages. No stage is fully autonomous and the latent effects of decisions ripple throughout the model. Each earlier stage creates a unique context for what transpires at later stages and later stages remain tethered to the former. For instance, previous research suggests that the number of warnings received during the reception stage goes on to shape what happens during later stages involving credibility and belief (Luo et al. 2015). Adding further weight to this argument, Ripberger et al. (2015b) notes that those unable to distinguish between watches and warnings tended to regard the False Alarm Rate (FAR) as higher than those who could, suggesting that outcomes at the understanding stage creates a context for what occurs later in the belief stage. What are the consequences of such a perspective?

To paraphrase the sociologist Merton, here it is important to remind ourselves that social policies and decisions have unintended consequences. We illustrate this point through an extended example. Warnings issued from multiple sources are more likely to be received (stage 1 of the model). Therefore, a sensible policy might be to simply issue as many warnings from as many sources as possible. However, this approach overlooks later stages in which individuals attempt to grasp the meaning of messages (i.e., understanding stage) and confirm the truth of messages (i.e., credibility stage). This is a problem because, as was demonstrated in previous sections, message consistency is key to warning credibility, comprehension, and the discouragement of excessive confirmation and milling behavior that might lead to delays in seeking shelter. For this reason, it is crucial at these stages that information received earlier in the reception stage be in a consistent and coherent form and with as few contradictions and ambiguities as possible. Here the contradiction becomes clear: the act of issuing multiple warnings, a smart idea if the intention is to reach as many people as possible, incidentally increases the chances of message inconsistency at later stages, either because the messages are themselves inconsistent or multiple sources of information are just naturally difficult to process into an overall picture of the risk one faces. Inconsistent, vague, or incomplete information among the various sources providing the information may cause recipients to either misunderstand the general meaning or doubt the credibility of intended warnings.

Directionality

To grasp the dynamic significance of directionality to the warning process, it must first be understood that warning response is not an individual process per se, but a complex process that unfolds over time, involving the interaction among and between individuals, groups, and communities (Quarantelli 1980b). Consequently, it follows that that different individuals, groups, and communities may be at different stages of the warning process at different times. Therefore, one individual or group involved in the process of determining whether protective action is necessary may contact another individual or group attempting to determine whether the warning information is true. This implies that later stages act on earlier stages precisely because the warning process is a dynamic and interactive social process in which different groups communicate and coordinate. For instance, someone who disregards a warning may phone someone else or another group still in the process of reception or negotiating an understanding of received warning information. The person who disregards the warning may pass off his or her disbelief to this group, stopping them at the stage of reception or understanding. Moreover, groups involved in evacuation may establish a norm (Turner and Killian 1987) by which others stuck at the credibility stage may be ushered into action. Overall, Nigg's (1982) non-linear model of risk communication illustrates the process well.

Demographic Variables Are Not Consistent Across Stages

Finally, the reader likely noticed a familiar refrain across all stages of the model—that is, the *inconsistency* with which demographic variables appear to influence the warning process. For instance, this is particularly true in the case of age; older persons are less likely to hear a warning message (Gruntfest 1997; Perry 1979; Turner et al. 1979); more likely to understand a warning message (Blanchard-Boehm 1998; Turner et al. 1979); less likely to believe a warning message (Hodge et al. 1981; Friedsam 1961); and more likely to evacuate (Aguirre et al. 1998). This is not necessarily a disadvantage, but rather a strength, of the warning model: it allows us to address more specifically the individual vulnerabilities each group faces during the warning process. Thus, it should not be assumed just because one demographic group has an advantage or disadvantage at one stage of the model, said groups will have similar advantages or disadvantages at all stages of the model.

Conclusion

It is often lamented among emergency managers that the community does not behave as intended when warnings are issued—they do not *do* what emergency managers intend or what is expected of them. In order to avoid psychologically reductive explanations, Chap. 3 has been an attempt to explain why this is the case, for if we diagnose better why individuals subvert our expectations regarding warning response, we can mold our technologies, policies, and approaches to better ensure that they do. Of course, if people immediately responded to risk as we believe they should, or as we intend them to, the development of warning systems would be straightforward and uncontroversial. But as demonstrated in the preceding Chapter, this is far from the case. On the contrary, public warning response is a complex, dynamic, and messy process, but it is one we can understand and harness in our development of interdisciplinary systems. In disputing the far-too-simple "stimulus-response" perspective on warning response, and introducing an empirically-grounded approach to understanding warning response, new possibilities for interdisciplinary research are introduced. We must view matters of lead time, understanding, credibility, and evaluations of the feasibility necessity of response as inseparable from the technical development of warning systems. And, if we are to truly innovate, our warning technologies must take these realities into account if we are to truly move forward in the development of remote sensing, radars, or any other technology intended to warn the public of impending hazards. In the next chapter, we now turn to another topic highly relevant to interdisciplinary research: institutional response.

Bibliography

Aguirre, B. E. (2004). "Los Desastres en Latinoamérica: Vulnerabilidad y Resistencia." Revista Mexicana de Sociologia. 66: 485–510.

Aguirre, B.E. (1988). "The Lack of Warnings Before the Saragosa Tornado." International Journal of Mass Emergencies and Disasters. 6: 65–74.

Aguirre, B.E. (1991). "Evacuation in Cancun During Hurricane Gilbert." International Journal of Mass Emergencies and Disasters. 9(1): 31–35.

Aguirre, B.E., Wenger, D.E., and Vigo, G. (1998). A Test of the Emergent Norm Theory of Collective Behavior. Sociological Forum. 13: 301–320.

Ahlborn, L and Franc, J.M. (2012). Tornado Hazard Communication Disparities among Spanish-Speaking Individuals in an English-Speaking Community. *Prehospital and Disaster Medicine*, 27(1), 98–102. doi: https://doi.org/10.1017/S1049023X12000015.

Anderson, J.W. (1968). Cultural Adaptation to Threatened Disasters. *Human Organization*, 27(4), 298-307.

Ashley, W.S., Krmenec, A., and Schwantes, R. (2008). Vulnerability Due to Nocturnal Tornadoes. *Weather and Forecasting*, 23(5), 795–807. doi: https://doi.org/10.1175/2008WAF2222132.1.

Atwood, L. E. and Major, A.M. (1998). "Exploring the "Cry Wolf" Hypothesis." *International Journal of Mass Emergencies and Disasters,* 16, 279-302.

Baker, E.J. (1979). Predicting Response to Hurricane Warnings - Reanalysis of Data from 4 Studies. *Mass Emergencies*, 4(1), 9–24.

Balluz, L., Schieve, L., Holmes, T., Kiezak, S., and Malilay, J. (2000). Predictors for people's response to a tornado warning: Arkansas, 1 March 1997. *Disasters*. 24(1): 71–77.

Baumann, D., and Sims, J.H. (1978). Flood Insurance: Some Determinants of Adoption. *Economic Geography*, 54(3), 189–96. doi: https://doi.org/10.2307/142833.

Beyth-Marom, R. (1982). How probable is probable? A numerical translation of verbal probability expressions. *Journal of Forecasting*, 1(3), 257–269.

Blanchard-Boehm, R.D. (1998). Understanding Public Response to Increased Risk from Natural Hazards: Application of the Hazards Risk Communication Framework. *International Journal of Mass Emergencies and Disasters*, 16(3), 247-278.

Bostrom, A., Morss, R., Lazo, J.K., Demuth, J., and Lazrus, H. (2018). Eyeing the Storm: How Residents of Coastal Florida See Hurricane Forecasts and Warnings. *International Journal of Disaster Risk Reduction*, 30, 105–19. doi: https://doi.org/10.1016/j.ijdrr.2018.02.027.

Breakwell, G.M. (2000). Risk communication: factors affecting impact. *British Medical Bulletin*, 56(1), 110–120.

Breznitz, S. (1984). Cry Wolf: The Psychology of False Alarms. Hillsdale, NJ: Lawrence: Erlbaum Associates Publishers.

Broad, K, Leiserowitz, A. Weinkle, J., and Steketee, M. (2007). Misinterpretations of the 'Cone of Uncertainty' in Florida during the 2004 Hurricane Season. *Bulletin of the American Meteorological Society*, 88(5), 651–68. doi: https://doi.org/10.1175/BAMS-88-5-651.

Byrnes, J. P., Miller, D. C., & Schafer, W. D. (1999). Gender differences in risk taking: a meta-analysis. Psychological Bulletin, 125, 367–383

Carter, Michael T. (1980). "Community Warning Systems: The Interface Between the Broadcast Media, Emergency Service Agencies and the National Weather Service." Pp. 214–228 in Disasters and the Mass Media. Washington, D.C.: Committee on Disasters and Mass Media, National Academy of Sciences.

Casteel, M.A. (2016). Communicating Increased Risk: An Empirical Investigation of the National Weather Service's Impact-Based Warnings. *Weather, Climate, and Society*, 8(3), 219–32. doi: https://doi.org/10.1175/WCAS-D-15-0044.1.

Casteel, M.A. (2018). An Empirical Assessment of Impact Based Tornado Warnings on Shelter in Place Decisions. *International Journal of Disaster Risk Reduction,* 30, 25–33. doi: https://doi.org/10.1016/j.ijdrr.2018.01.036.

Clifford, R.A. (1956). "The Rio Grande Flood: A Comparative Study of Border Communities." Washington, D.C.: National Research Council, National Academy of Sciences.

Cola, R.M. 1996. "Responses of Pampanga Households to Lahar Warnings: Lessons from Two Villages in the Pasig-Potrero River Watershed." Pp. 141–149 in Fire and Mud: Eruptions and Lahars of Mount Pinatubo, Philippines, edited by C. G. Newhall and R. S. Punongbayan. Seattle, WA: University of Washington Press.

Croson, Rachel, and Uri Gneezy. 2009. "Gender Differences in Preferences." Journal of Economic Literature, 47 (2), 448–74.

Crowe, A. (2010). When Status Quo Becomes Obsolete: The Changing Use of Outdoor Warning Sirens. *Journal of Homeland Security and Emergency Management,* 7(1), 1–5 doi: https://doi.org/10.2202/1547-7355.1758.

Dash, N. (2002). Decision-Making under Extreme Uncertainty: Rethinking Hazard Related Perceptions and Action. *FIU Electronic Theses and Dissertations.* doi: https://doi.org/10.25148/etd.FI14062203.

Dash, N. and Gladwin, H. (2007). Evacuation Decision Making and Behavioral Responses: Individual and Houeshold. *Natural Hazards Review*, 8(3), 69–77.

Demerath, Nicholas. (1957). Some General Propositions: An Interpretive Summary. *Human Organization,* 16 (Summer), 28–29

Parker, D.J., Tunstall, S.M. and McCarthy, S. (2007). New insights into the benefits of flood warnings: Results from a household survey in England and Wales, Environmental Hazards, 7(3), 193–210,

Deschêne, A, Stahl, C., Hansen, J., and Wänke, M. (2010). The Truth About the Truth: A Meta-Analytic Review of the Truth Effect. *Personality and Social Psychology Review*, 14(2), 238–57. doi: https://doi.org/10.1177/1088868309352251.

Diggory, J.C. (1956). Some Consequences of Proximity to a Disease Threat. *Sociometry*, 19(1), 47–53.

Donner, W. (2007). The Political Ecology of Disaster: An Analysis of Factors Influencing U.S. Tornado Fatalities and Injuries, 1998-2000. *Demography*, 44(3), 669–685.

Donner, W.R., and Lavariega-Montforti, J. (2018). Ethnicity, Income, and Disaster Preparedness in Deep South Texas, United States, Disasters 42(4), 719–33. doi: https://doi.org/10.1111/disa.12277

Donner, W. and Rodríguez, H. (2008). "Population Composition, Migration and Inequality: The Influence of Demographic Changes on Disaster Risk and Vulnerability, Social Forces." 87(2): 1089–1114, https://doi.org/10.1353/sof.0.0141

Donner, W., Rodriguez, H. and Diaz, W. (2012). Tornado Warnings in Three Southern States: A Qualitative Analysis of Public Response Patterns. *Journal of Homeland Security and Emergency Management*, 9(2), 1–19.

Drabek T.E. (1994). "Disaster Evacuation and the Tourist Industry." Boulder, CO: Natural Hazards Center.

Drabek, T.E. (1969). Social Processes in Disaster: Family Evacuation. *Social Problems*, 16(3), 336–49. doi: https://doi.org/10.2307/799667.

Drabek, T.E. (2001). Disaster Warning and Evacuation Responses by Private Business Employees. *Disasters*, 25(1), 76–94.

Drabek, T.E. (2000). The social factors that constrain human responses to flood warnings. In: Parker, D.J. (Ed.), Floods, vol. 1. Routledge, London.

Drost, R, Casteel, M., Libarkin, J., Thomas, S., and Meister, M. (2016). Severe Weather Warning Communication: Factors Impacting Audience Attention and Retention of Information during Tornado Warnings. *Weather, Climate, and Society*, 8(4), 361–72.

Eastern Research Group. (2018). National Weather Service Hazard Simplification: Public Survey. Final Report. Lexington, MA.

Edwards, M.L. (1993). Social Location and Self-Protective Behavior: Implications for Earthquake Preparedness. *International Journal of Mass Emergencies and Disasters*, 11(3), 293–303.

Ewald, R. and Guyer, J. (2002). The Ideal Lead Time For Tornado Warnings - A Look From The Customer's Perspective Customer's Perspective Presented at American Meteorological Society, The 21st Conf. on Severe Local Storms and 19th Conf. on WeatheZAnalysis and Forecasting/ 15th Conf. on Numerical Weather Prediction, August 16, 2002.

Farley, J. E. , Barlow, H. D. , Finkelstein, M. S. , & Riley, L. (1993). Earthquake hysteria, before and after: A survey and follow-up on public response to the Browning forecast. International Journal of Mass Emergencies and Disasters, 11, 305–322.

Farmer, A.K., DeYoung, S.E. and Wachtendorf, T. (2016). Pets and Evacuation: An Ongoing Challenge in Disasters. *Journal of Homeland Security and Emergency Management*, 13(4), 1–13.

Fazio, L.K., Rand, D.G., and Pennycook, G. (2019). "Repetition Increases Perceived Truth Equally for Plausible and Implausible Statements" *Psychonomic Bulletin & Review,* 26, 1705–10.

Fischer, HW., Stine, G., Stoker, B., Trowbridge, M., and Drain, E. (1995). "Evacuation Behaviour: Why Do Some Evacuate, While Others Do Not? A Case Study of the Ephrata, Pennsylvania (USA) Evacuation." *Disaster Prevention and Management: An International Journal*, 4(4), 30–36. doi: https://doi.org/10.1108/09653569510093414.

Flynn, C.B. (1979). "Three Mile Island Telephone Survey: Preliminary Report on Procedures and Findings." Tempe, AZ: Mountain West Research.

Flynn, J., Slovic, P., and Mertz, C.K. (1993). The Nevada Initiative: A Risk Communication Fiasco. *Risk Analysis*, 13(5), 497–502. doi: https://doi.org/10.1111/j.1539-6924.1993.tb00007.x.

Foster, H.D. (1980), Disaster Planning: The Preservation of Life and Property, Springer, New York, NY.

Fothergill, A. and Peek, L. (2004). Poverty and Disasters in the United States: A Review of Recent Sociological Findings. Natural Hazards, 32, 89–110. doi: https://doi.org/10.1023/B:NHAZ.0000026792.76181.d9.

Friedsam, H. J. (1961). "Reactions of Older Persons to Disaster-Caused Losses: A Hypothesis of Relative Deprivation." *Gerontologist*, 1, 34–37.

Fritz, C. E., and Marks, E.S. (1954). The NORC Studies of Human Behavior in Disaster. *Journal of Social Issues,* 10(3), 26–41. doi: https://doi.org/10.1111/j.1540-4560.1954.tb01996.x.

Fritz, C.E. (1961). Disasters. Pp. 651–694 in Contemporary Social Problems, edited by R.K. Merton and R. A. Nisbet. New York, NY: Harcourt.

Fritz, Charles E. (1957). Disasters Compared in Six American Communities. *Human Organization*, 16, 6–9.

Fritz Institute. (2006). *Hurricane Katrina: Perceptions of the Affected*. Report. Retrieved 6/26/2022 (https://reliefweb.int/report/united-states-america/hurricane-katrina-perceptions-affected).

Gladwin, H. and Peacock, W.G. (1997). Warning and evacuation: a night for hard houses. In: *Hurricane Andrew: Ethnicity, Gender and the Sociology of Disasters*, Pp. 52–72, edited by Peacock, W.G., Gladwin, H., and Morrow, B.H. 1997. London: Routledge.

Gladwin, H., Lazo, J.K., Morrow, B.H., Peacock, W.G., and Willoughby, H. (2007). Social Science Research Needs for the Hurricane Forecast and Warning System. *Natural Hazards Review, 8*(3), 87–95.

Godfrey, C., Wolf, P., Goldsbury, M., Caudill, J., and Wedig, D. (2011). An Evaluation of Connective Warning Utilization by the general public. Retrieved December 26, 2020 (/ paper/1.3-AN-EVALUATION-OF-CONVECTIVE-WARNING-UTILIZATION-Caudill-Wedig/f38c344ce3077c0efbd9d7423237b327f43fb597).

Goldsteen, R., and Schorr, J. (1982). The Long-Term Impact of a Man-Made Disaster: An Examination of a Small Town in the Aftermath of the Three Mile Island Nuclear Reactor Accident. *Disasters*, 6(1), 50–59.

Grazulis, T. P. (2001). The Tornado: Nature's Ultimate Windstorm. Norman, OK: University of Oklahoma Press.

Griffin, R.J., Neuwirth, K., Dunwoody, S., and Giese, J. (2004). Information Sufficiency and Risk Communication. *Media Psychology*, 6(1), 23–61.

Gruntfest, E. (1997). "Warning Dissemination and Response with Short Lead Times." Pp. 191- 202 in Flood Hazard Management: British and International Perspectives, edited by J. Handmer. Norwich, UK: GEO Books.

Guillot, S., Jarvis, P., Powell, T., and Kenkre, J. (2020). Knowledge, Experience and Preparedness: Factors Influencing Citizen Decision-Making in Severe Weather Situations. *International Journal of Emergency Management*, 16(1), 60–77.

Gutteling, J.M., Terpstra T., and Kerstholt, J.H. (2018). Citizens' Adaptive or Avoiding Behavioral Response to an Emergency Message on Their Mobile Phone. *Journal of Risk Research*, 21(12), 1579–91.

Haas, J.E., Cochrane, H. & Eddy, D. (1977). Consequences of a cyclone for a small city. *Ekistics*, 44, 45–51.

Hammer, B. and T.W. Schmidlin. (2002). Response to Warnings During the 3 May 1999 Oklahoma City Tornado: Reasons and Relative Injury Rates. *Weather and Forecasting*, 17, 577–581.

Hansson, R.O., D. Noulles, and S.J Bellovich. (1982). "Knowledge, Warning and Stress." Environment and Behavior, 14 (2), 171–185.

Harris, C.R., Jenkins, M., and Glaser, D. (2006). Gender Differences in Risk Assessment: Why do Women Take Fewer Risks than Men? *Judgment and Decision Making*, 1(1), 48–63.

Hodge, David, Virginia Sharp, and Marion Marts. (1981). "Contemporary Responses to Volcanism: Case Studies From the Cascades and Hawaii." Pp. 221–248 in Volcanic Activity and Human Ecology, edited by P. D. Sheets and D. K. Grayson. Elsevier.

Hoekstra, S., Klockow, K., Riley, R., Brotzge, J., Brooks, H., & Erickson, S. (2011). A Preliminary Look at the Social Perspective of Warn-on-Forecast: Preferred Tornado Warning Lead Time and the General Public's Perceptions of Weather Risks. *Weather, Climate, and Society*, 3(2), 128–140

Lindell, Michael & Lu, Jing-Chein & Prater, Carla. (2005). Household Decision Making and Evacuation in Response to Hurricane Lili. Natural Hazards Review. 6. https://doi.org/10.1061/(ASCE)1527-6988(2005)6:4(171).

Houts, P. S., et al. (1984). Protective Action Decision Model Applied to Evacuation During the Three Mile Island Crisis. *International Journal of Mass Emergencies and Disasters*, 2(1), 27–39.

Hu, X., Zhang, X., & Wei, J. (2019). Public Attention to Natural Hazard Warnings on Social Media in China, Weather, Climate, and Society, 11(1), 183–197.

Huang, S.K., Lindell M.K., Prater, C.S. (2016). Who Leaves and Who Stays? A Review and Statistical Meta-Analysis of Hurricane Evacuation Studies. *Environment and Behavior*, 48(8), 991–1029.

Lee, H., Hong, W., and Lee, Y. (2019). Experimental study on the influence of water depth on the evacuation speed of elderly people in flood conditions, *International Journal of Disaster Risk Reduction*, 39(6B):101198.

Janda, R.J., et al. (1996). Assessment and Response to Lahar Hazard Around Mount Pinatubo, 1991 to 1993. Pp. 107- 139 in *Fire and Mud: Eruptions and Lahars of Mount Pinatubo*, Philippines, edited by C. G. Newhall and R. S. Punongbayan. Seattle, WA:

Jauernic, S.T. and Van Den Broeke, M.S. (2016). Perceptions of Tornadoes, Tornado Risk, and Tornado Safety Actions and Their Effects on Warning Response among Nebraska Undergraduates. *Natural Hazards*, 80(1), 329–50. doi: https://doi.org/10.1007/s11069-015-1970-9.

Joan Shorenstein Center on the Press, Politics and Public Policy (2007). Young people and the news. www.ksg.harvard.edu/presspol/research_publications/papers/research_papers/R29.pdf, Accessed July 18, 20019.

Jon, I., Huang, S., and Lindell, M.K. (2019). Perceptions and Expected Immediate Reactions to Severe Storm Displays. *Risk Analysis*, 39(1), 274–90.

Joslyn, S., Nadav-Greenberg, L., & Nichols, R. M. (2009). Probability of Precipitation: Assessment and Enhancement of End-User Understanding. *Bulletin of the American Meteorological Society*, *90*(2), 185–194.

Kim, J. (2019). "Why People Remain Inactive during a Crisis: Interpreting and Dealing with a Crisis within a Broader Social Context." *Health, Risk & Society*, 21(1–2), 74–88.

Kirschenbaum, A. (1992). Warning and Evacuation During a Mass Disaster: A Multivariate Decision-Making Model. *International Journal of Mass Emergencies and Disasters*, 10(1), 91–114.

Klinenberg, E. (2001). "Dying Alone: The Social Production of Urban Isolation." *Ethnography*, 2(4), 501–31. doi: https://doi.org/10.1177/14661380122231019.

Klockow, K.E., Peppler, R.A. and McPherson, R.A. (2014). Tornado folk science in Alabama and Mississippi in the 27 April 2011 tornado outbreak. *GeoJournal*, 79, 791–804.

Lachman, R., Tatsuoka, M, and Bonk, W. (1961). Human Behavior during the Tsunami of 1960. *Science*. 133(3462): 1405–1409.

Laidlaw, E., (2010), The Controversy Over Outdoor Warning Systems, *Weatherwise*. 63, 1.

Lardry, T. and Rogers, G. (1982). Warning Confirmation and Dissemination. Pittsburgh, PA: Center for Social and Urban Research, University of Pittsburgh.

Lasker, R. D. (2004). Redefining Readiness: Terrorism Planning through the Eyes of the Public. *Digital Collections - National Library of Medicine*. https://collections.nlm.nih.gov/catalog/nlm:nlmuid-101233378-pdf. Retrieved December 26, 2020.

Lenhardt, E. et al. (2020). How Likely Is That Chance of Thunderstorms? A Study of How National Weather Service Forecast Offices Use Words of Estimative Probability and What They Mean to the Public." *Journal of Operational Meteorology*, 8(5), 64–78.

Levine, Robert N. (2008). *A Geography Of Time: On Tempo, Culture, And The Pace Of Life*. NY: Basic Books.

Li, J. (1991). Social Responses to the Tangshan Earthquake Prediction. Presented at the UCLA International Conference on the Impact of Natural Disasters. Los Angeles, CA: University of California-Los Angeles.

Lim, J. R., Liu, B. F., & Egnoto, M. (2019). Cry Wolf Effect? Evaluating the Impact of False Alarms on Public Responses to Tornado Alerts in the Southeastern United States *Weather, Climate, and Society*, 11(3), 549–563.

Lindell, M.K. and Perry, R.W. (1983). Nuclear Power Plant Emergency Warnings: How Would the Public Respond? *Nuclear News*, February, 49–57.

Lindell, M.K. and R.W. Perry. 1992. Behavioral Foundations of Community Emergency Planning. Bristol, PA: Hemisphere Publishing Corporation.

Lindell, M.K. and R.W. Perry. (2004). Communicating Environmental Risk in Multiethnic Communities. Thousand Oaks, CA: Sage.

Lindell, M.K., Huang, S.K., Wei, H.L., and Samuelson, C. (2016). Perceptions and expected immediate reactions to tornado warning polygons. *Natural Hazards*, 80(1), 683–707

Lindell, M.K. and Perry, R.W. (1987). Warning Mechanisms in Emergency Response Systems. *International Journal of Mass Emergencies and Disasters*, 5(2), 137–153.

López-Vázquez, E. and Marván, M. (2012). Volcanic Risk Perception, Locus of Control, Stress and Coping Responses of People Living Near the Popocatepetl Volcano in Mexico. *Journal of Risk Analysis and Crisis Response*. 2(1), 3–12. https://doi.org/10.2991/ijcis.2012.2.1.1.

Losee, J.E., and Joslyn, S. (2018). The Need to Trust: How Features of the Forecasted Weather Influence Forecast Trust. *International Journal of Disaster Risk Reduction*, 30, 95–104.

Luo, J., Cong, Z., and Liang, D. (2015). Number of Warning Information Sources and Decision Making During Tornadoes. *American Journal of Preventive Medicine*, 48(3), 334–37.

Mack, R.W. and Baker, G.W. (1961). The Occasion Instant: The Structure of Social Responses to Repeated Air Raid Warnings. National Research Council, Publication No. 945, Washington, D.C.: National Academy of Sciences.

Maercker, A., Ben-Ezra, M., Esparza, O. A., and Augsburger, M. (2019). Fatalism as a traditional cultural belief potentially relevant to trauma sequelae: Measurement equivalence, extent and associations in six countries. *European Journal of Psychotraumatology*, 10(1), 1657371.

Maldonado, A., Collins, T.W., and Grineski, S.E. (2016). "Hispanic Immigrants' Vulnerabilities to Flood and Hurricane Hazards in Two United States Metropolitan Areas." *Geographical Review*, 106(1), 109–35.

Mason, J.B., and Senkbeil, J.C. (2014). "Implications of the 2011 Tuscaloosa EF4 Tornado for Shelter and Refuge Decisions." *Natural Hazards*, 74(2), 1021–41.

McDavid, J. and M. Marai. (1968). *Social Psychology*. New York, NY: Harper and Row.

Meléndez-Landaverde, E.R., Werner, M., and Verkade, J. (2020). Exploring Protective Decision-Making in the Context of Impact-Based Flood Warnings. *Journal of Flood Risk Management*, 13(1), e12587.

Meyer M.A. (2018). Social Capital, in Disaster Research. In: Rodríguez H., Donner W., Trainor J. (eds) Handbook of Disaster Research. Handbooks of Sociology and Social Research. Cham., Switzerland: Springer.

Mileti, D. S. (1975). Natural hazard warning systems in the United States: A research assessment. Washington, D.C.: University of Colorado.

Mileti, D. 1999. Disasters by Design: A Reassessment of Natural Hazards in the United States. Washington, D.C.: Joseph Henry Press.

Mileti, D. S., and J. H. Sorensen. (1990). *Communication of Emergency Public Warnings: A Social Science Perspective and State-of-the-Art Assessment*. ORNL-6609. Oak Ridge National Lab., TN (USA).

Mileti, D.S. and C. Fitzpatrick. (1993). Great Earthquake Experiment: Risk Communication and Public Action. Boulder, CO: Westview Press.

Mileti, D.S. and Darlington, J.D. (1995). Societal Response to Revised Earthquake Probabilities in the San Francisco Bay Area. *International Journal of Mass Emergencies and Disasters*, 13(2), 119–145.

Mileti, D.S. and J.D. Darlington. 1997. "The Role of Searching in Shaping Reactions to Earthquake Risk Information." *Social Problems*. 44: 89-103.

Mileti, D.S. and Fitzpatrick, C. (1992). Causal Sequence of Risk Communication in the Parkfield Earthquake Prediction Experiment. *Risk Analysis*, 12 (3), 393–400.

Miran, S.M., Ling, C. and James, J. (2020). People's Thresholds of Decision-Making against a Tornado Threat Using Dynamic Probabilistic Hazard Information. *International Journal of Disaster Risk Reduction*, 42, 101345.

Mitchem, J.D. (2003). An Analysis of the September 20, 2002, Indianapolis Tornado: Public Response to a Tornado Warning and Damage Assessment Difficulties. Natural Hazards Research and Applications Information Center, University of Colorado.

Molinari, D. and Handmer, J. (2011). "A behavioural model for quantifying flood warning effectiveness." *Journal of Flood Risk Management*, 4(1), 23–32.

Moore, H.E. (1958). Tornadoes Over Texas. Austin, TX: University of Texas Press.

Morss, R and Hayden, M. (2010). Storm Surge and 'Certain Death': Interviews with Texas Coastal Residents Following Hurricane Ike. Weather and Forecasting, 2, 174-189.

Morss, R.E., Mulder, K., Lazo, J.K. and Demuth, J.L. (2016). How Do People Perceive, Understand, and Anticipate Responding to Flash Flood Risks and Warnings? Results from a Public Survey in Boulder, Colorado, USA. Journal of Hydrology, 541, 649–64. doi: https://doi.org/10.1016/j.jhydrol.2015.11.047

Nigg, J.M. (1982). "Communication Under Conditions of Uncertainty: Understanding Earthquake Forecasting." *Journal of Communication*, 32(1), 27–36.

Nigg, J. M. (1987). Communication and behavior: Organizational and individual response to warnings. In Dynes, R. R., DeMarchi, B., Pelanda, C. (Eds.), Sociology of disasters (pp. 103–117). Milan, Italy: Franco Angeli Libri.

Nomura S., Blangiardo M., Tsubokura M., Nishikawa Y., Gilmour S., Kami M., and Hodgson S. (2016). "Post-Nuclear Disaster Evacuation and Survival amongst Elderly People in Fukushima: A Comparative Analysis between Evacuees and Non-Evacuees." *Preventive Medicine*, 82, 77–82.

Parker, J.A., Whitmer, D.E., and Sims, V.K. (2018). Warnings for Hurricane Irma: Trust of Warning Type and Perceptions of Self-Efficacy and Susceptibility. *Proceedings of the Human Factors and Ergonomics Society Annual Meeting*, 62(1), 1368–72.

Patterson, T. (2007) Young People and the News. Report. Harvard University: Joan Shorenstein Center on the Press, Politics and Public Policy.

Paul, B., V. Brock, S. Csiki, and L. Emerson. (2003). Public response to tornado warnings: A comparative study of the May 4, 2003, tornados in Kansas, Missouri, and Tennessee. Natural Hazards Center. Quick Response Research Report #165, Natural Hazards Research and Applications Information Center, University of Colorado

Peacock, W.G., Morrow, B.H., and Gladwin, H (eds.). (1997). *Hurricane Andrew: Ethnicity, Gender, and the Sociology of Disasters*. London: Routledge.

Pearson, R.G. and M.G. Joost. (1983). "Egress Behavior Response Times of Handicapped and Elderly Subjects to Simulated Residential Fire Situations." U.S. Department of Commerce and U.S. Department of Health and Human Services.

Peek, L. (2008). Children and Disasters: Understanding Vulnerability, Developing Capacities, and Promoting Resilience — An Introduction. Children, Youth and Environments, 18(1), 1-29.

Pennington, R.H., Wilcox, D., and Grover, V. (2003). "The Role of System Trust in Business-to-Consumer Transactions." *Journal of Management Information Systems*, 20(3), 197–226.

Perry, R. W., Lindell, M.K. and Tierney, K. 2001. *Facing the Unexpected: Disaster Preparedness and Response in the United States*. Joseph Henry Press.

Perry, R.W. (1979). "Evacuation Decision-Making in Natural Disasters." *Mass Emergencies*, 4, 25–38.

Perry, R.W. 1981. "Citizen Evacuation in Response to Nuclear and Non-Nuclear Threats." Seattle, WA: Battelle Human Affairs Research Center.

Perry, R.W. (1983). Population evacuation in volcanic eruptions, floods, and nuclear power plant accidents: Some elementary comparisons. *Journal of Community Psychology*, 11(1), 36–47.

Perry, R.W. 1987. "Disaster Preparedness and Response Among Minority Citizens." Pp.135–151 in Sociology of Disasters, edited by R. R. Dynes, B., DeMarchi and C. Pelanda. Milan, Italy: Franco Angeli Libri.

Perry, R.W. and H. Hirose. (1991). Volcano Management in the United States and Japan. Greenwich, CT: Jai Press.

Perry, R.W. and Green, M.R. (1982). "The Role of Ethnicity in the Emergency Decision-Making Process." Sociological Inquiry 52(4):309-334

Perry, R.W. and M.K. Lindell. 1986. "Twentieth-Century Volcanicity at Mt. St. Helens: The Routinization of Life Near an Active Volcano." Tempe, AZ: School of Public Affairs, Arizona State University.

Perry, R.W., Lindell, M.K. and Greene, M.R. (1981). *Evacuation Planning in Emergency Management*. Lexington, MA: Heath Lexington Books.

Perry, R.W., Lindell, M.K., and Greene, M.R. (1982). Threat Perception and Public Response to Volcano Hazard. *The Journal of Social Psychology*, 116(2), 199–204.

Perry, R.W., M.K. Lindell, and M.R. Greene. (1980). The Implications of Natural Perspectives, edited by J. Handmer. Norwich, UK: GEO Books.

Peters, R.G., Covello, V.T. and McCallum, D.B. (1997). The Determinants of Trust and Credibility in Environmental Risk Communication: An Empirical Study. Risk Analysis, 17(1), 43–54.

Pfister, N. (2002). Community Response to Flood Warnings: The Case of an Evacuation from Grafton, March 2001. *The Australian Journal of Emergency Management*, 17(2):19–29.

Pickles, et al. (2020). COVID-19: Beliefs in Misinformation in the Australian Community. *MedRxiv* 2020.08.04.20168583.
Ponting, J.R. (1974). It Can't Happen Here: A Pedagogical Look at Community Coordination in Response to a Toxic Gas Leak. Emergency Planning Digest, 1, 8–13.
Potter, S.H. et al. (2018). The Influence of Impact-Based Severe Weather Warnings on Risk Perceptions and Intended Protective Actions. *International Journal of Disaster Risk Reduction*, 30, 34–43.
Powell, S. (2008). Communicating Weather Information to the Public: People's Reactions and Understandings of Weather Information and Terminology. 3rd Symposium on Policy and Socioeconomic Impacts, AMS Annual Meeting, New Orleans, LA.
Quarantelli, E. L. (1980a). Evacuation Behavior and Problems: Findings and Implications from the Research Literature. Columbus, OH: Disaster Research Center, Ohio State University.
Quarantelli, E.L. (1980b). "Some Research Emphases for Studies on Mass Communication Systems and Disasters." In 239-299 in Proceedings of the Committee on Disaster and the Mass Media Workshop. Washington, D.C.: National Academy of Sciences.
Quarantelli, E.L. (1991). Lessons from research: findings on mass communication system behavir in the pre, trans, and postimpact period of disasters. PRELIMINARY PAPER #160, Disaster Research Center, University of Delaware, Newark, DE.
Ramey-Smith, A.M. and J.V. Fletcher. (1978). Group Homes for the Developmentally Disabled. Center for Consumer Product Technology, National Bureau of Standards, Washington, D.-C. During Hurricane Eloise. Report 51, Social Science Research Center, Mississippi State University, Starksville, MI.
Ripberger, J. T., Krocak, M. J., Wehde, W. W., Allan, J. N., Silva, C., & Jenkins-Smith, H. (2019). Measuring Tornado Warning Reception, Comprehension, and Response in the United States, *Weather, Climate, and Society, 11*(4), 863–880.
Ripberger, J. T., Silva, C., Jenkins-Smith, H.C., and James, M. (2015a). "The Influence of Consequence-Based Messages on Public Responses to Tornado Warnings." *Bulletin of the American Meteorological Society*, 96(4), 577–90.
Ripberger, J.T., Silva, C., Jenkins-Smith, H.C., Carlson, D.E., James, M., and Herron, K.G. (2015b). "False Alarms and Missed Events: The Impact and Origins of Perceived Inaccuracy in Tornado Warning Systems." *Risk Analysis*, 35(1), 44–56.
Rodríguez, H., Díaz, W., J. Santos, and B. E. Aguirre. 2006. "Communicating Risk and Uncertainty: Science, Technology, and Disasters at the Crossroads." Pp. 476-488 in Handbook of Disaster Research, edited by H. Rodriguez, E. Quarantelli, and R. Dynes. New York: Springer
Rogers, G.O. and Nehnevajsa, J. (1984). "Crisis Conditions." Pittsburgh, PA: Center for Social and Urban Research, University of Pittsburgh.
Rogers, G. O., and Nehnevajsa, J. (1987). Warning Human Populations of Technological Hazard. *American Nuclear Society*, 357–62.
Roulston, M.S., and Kaplan, T.R. (2009). A Laboratory-Based Study of Understanding of Uncertainty in 5-Day Site-Specific Temperature Forecasts. *Meteorological Applications*, 16(2), 237–44.
Saarinen, T.F., Baker, V.R., Durrenberger, R. and Maddock, T. (1984). The Tucson, Arizona Flood of October 1983. National Research Council, Committee on Natural Disasters. Washington, D.C.: National Academy Press.
Bhavaraju, S.K.T., Beyney, C., and Nicholson, C. (2019). Quantitative analysis of social media sensitivity to natural disasters. *International Journal of Disaster Risk Reduction*, 39, 101251.
Scanlon, J. and Frizzell, A. (1979). Old Theories Don't Apply: Implications of Communications in Crises. *Disasters*, 3(3), 315–19.
Schumann, R. L., Ash, K.D., and Bowser, G.C. (2018). Tornado Warning Perception and Response: Integrating the Roles of Visual Design, Demographics, and Hazard Experience. *Risk Analysis*, 38(2), 311–32.

Senkbeil, J.C., Scott, D.A., Guinazu-Walker, P.A., and Rockman, M.S. (2014). Ethnic and Racial Differences in Tornado Hazard Perception, Preparedness, and Shelter Lead Time in Tuscaloosa. *The Professional Geographer*, 66(4), 610–20.

Sherman-Morris, K., Pechacek, T., Griffin, D.J., and Senkbeil, J. (2020). Tornado Warning Awareness, Information Needs and the Barriers to Protective Action of Individuals Who Are Blind. *International Journal of Disaster Risk Reduction*, 50, 101709.

Sherman-Morris, Kathleen. (2005). Tornadoes, Television and Trust—A Closer Look at the Influence of the Local Weathercaster during Severe Weather. *Global Environmental Change Part B: Environmental Hazards*, 6(4), 201–10.

Sherman-Morris, Kathleen. (2010. Tornado Warning Dissemination and Response at a University Campus. *Natural Hazards*, 52(3), 623–38.

Shuhei, N., et al. (2016). Post-nuclear disaster evacuation and survival amongst elderly people in Fukushima: A comparative analysis between evacuees and non-evacuees. *Preventive Medicine*, 82, 77–82.

Siegrist, M. and Cvetkovich, G. (2000). Perception of Hazards: The Role of Social Trust and Knowledge. *Risk Analysis*, 20(5), 713–20.

Silver, Amber, and Jean Andrey. (2014). "The Influence of Previous Disaster Experience and Sociodemographics on Protective Behaviors during Two Successive Tornado Events." *Weather, Climate, and Society*, 6(1), 91–103.

Silver, Amber. (2015). "Watch or Warning? Perceptions, Preferences, and Usage of Forecast Information by Members of the Canadian Public." *Meteorological Applications*, 22(2), 248–55.

Simmons, K. M. and Sutter, D. (2009): False Alarms, Tornado Warnings and Tornado Casualties. *Weather, Climate and Society*, 1(1), 38–53.

Simmons, K. M. and Sutter, D. (2008). Tornado Warnings, Lead Times, and Tornado Casualties: An Empirical Investigation. *Weather and Forecasting*, 23(2), 246–258.

Sivle, A.D., Kolstø, S.D., Hansen, P., and Kristiansen, J. (2014). How Do Laypeople Evaluate the Degree of Certainty in a Weather Report? A Case Study of the Use of the Web Service. *Weather, Climate, and Society*, 6(3), 399–412.

Slovic, P. (1993). Perceived Risk, Trust, and Democracy. *Risk Analysis*, 13(6), 675–82.

Smith, Keith. (2001). *Environmental Hazards: Assessing Risk and Reducing Disaster*. Psychology Press.

Sorensen, J.H. (1982). Evaluation of Emergency Warning System at Ft. St. Vrain Nuclear Power Plant. Oak Ridge, TN: Oak Ridge National Laboratory.

Sorensen, John H., and Philip J. Gersmehl. (1980). "Volcanic Hazard Warning System: Persistence and Transferability." *Environmental Management*, 4(2), 125–36.

Sorensen, J. and Mileti D. (1988). Warning and evacuation: answering some basic questions. *Industrial Crisis Quarterly*, 2, 195–209.

Spittal, M., McClure, J., Siegert, R., and and Walkey, F. (2008). "Predictors of Two Types of Earthquake Preparation." *Environment and Behavior*, 40(6), 798–817.

Stokoe, R.M. (2016). "Putting People at the Centre of Tornado Warnings: How Perception Analysis Can Cut Fatalities." *International Journal of Disaster Risk Reduction*, 17, 137–53. doi: https://doi.org/10.1016/j.ijdrr.2016.04.004.

Tierney, K. (1987). "Chemical Emergencies, Offsite Exposures and Organizational Response." Quick Response Report No. 21, Natural Hazards Research and Applications Center, Institute of Behavioral Science, University of Colorado, Boulder, CO.

Trainor, J.E, Nagele, D., Philips, B., and Scott, B. (2015). Tornadoes, Social Science, and the False Alarm Effect. *American Meteorological Society*, 7(4), 333–52.

Turner, R.H. (1976). Earthquake Prediction and Public Policy: Distillations from a National Academy of Sciences Report. *Mass Emergencies* 1:179-202.

Turner, R.H. and Killian, L.M. (1987). *Collective Behavior*. Third Edition. Englewood Cliffs, NJ: Prentice-Hall, Inc.

Turner, R.H., J.M. Nigg, D.H. Paz, and B.S. Young. (1979). "Earthquake Threat: The Human Response in Southern California." Los Angeles, CA: Institute for Social Science Research, University of California, Los Angeles.

US EPA, OAR. (2014). "The Inside Story: A Guide to Indoor Air Quality." *US EPA*. Retrieved December 26, 2020 (https://www.epa.gov/indoor-air-quality-iaq/inside-story-guide-indoor-air-quality).

Voght, B.M. (1991). Issues in Nursing Home Evacuation. *International Journal of Mass Emergencies and Disasters*, 9(2), 247–267.

Mihunov, V., Lam, N., Zou, L., Wang, Z., Wang, K. (2020). Use of Twitter in disaster rescue: lessons learned from Hurricane Harvey. *International Journal of Digital Earth*, 13(12), 1454–1466,

Walters, J.E., Mason, L.R., Ellis, K., and Winchester, B. (2020). Staying Safe in a Tornado: A Qualitative Inquiry into Public Knowledge, Access, and Response to Tornado Warnings. *Weather and Forecasting*, 35, 67–81.

Weinstein, N. D., & Klein, W. M. (1995). Resistance of personal risk perceptions to debiasing interventions. Health Psychology, 14(2), 132–140. https://doi.org/10.1037/0278-6133.14.2.132

Whittaker, J., Taylor, M., and Bearman, C. (2020). Why Don't Bushfire Warnings Work as Intended? Responses to Official Warnings during Bushfires in New South Wales, Australia. *International Journal of Disaster Risk Reduction*, 45, 101476.

Williams, C., Miller, P., Black, A. and Knox, J. (2017). Throwing Caution to the Wind: National Weather Service Wind Products as Perceived by a Weather-Salient Sample. *Journal of Operational Meteorology*, 5(9), 103–20.

Windham G.O., Posey E.I., Ross P.J., Spencer. B. (1977). Reaction to storm threat during hurricane Eloise. Social Science Research Center, Mississippi State University, Starkville, MS.

Wintle, B., H. Fraser, B. Wills, A. Nicholson, and F. Fidler. (2019). "Verbal probabilities: *Very likely* to be *somewhat* more confusing than numbers." *PLoS ONE*, 14(4), 1–18.

Xie, X.F., Wang, M., Zhang, R.G., Li, J. and Yu, Q.Y. (2011), The Role of Emotions in Risk Communication. *Risk Analysis*, 31, 450–465.

Zhao, M, Rosoff, H., and John, R.S. (2019). Media Disaster Reporting Effects on Public Risk Perception and Response to Escalating Tornado Warnings: A Natural Experiment. *Risk Analysis*, 39(3), 535–52.

Wang, Z, Lam, N., Obradovich, N., Ye, X. (2019). Are vulnerable communities digitally left behind in social responses to natural disasters? An evidence from Hurricane Sandy with Twitter data. *Applied Geography*, Volume 108(1), 1–8,

Zunkel, P. (2015). The Spatial Extent and Coverage of Tornado Sirens in San Marcos, Texas. *Applied Geography*, 60, 308–12.

Chapter 4
Institutional Warning Response: Emergency and Scientific Institutions

As was seen in Chap. 2, the humble beginnings of meteorology have since evolved into a vast, complex network of innovation out of which has emerged unprecedented development in prediction and detection technologies. To accommodate this growing complexity, a similarly complex network of bureaucratic institutions was required and naturally arose, populated by specialized roles (e.g., forecasters, emergency managers), groups (e.g., storm spotters), and organizations (e.g. NOAA, NWS, SPC). Whereas the past saw the lone meteorologist making predictions with minimal information and rudimentary—though at times highly innovative—instrumentation, today highly-trained and well-coordinated forecasting offices take advantage of remote sensing technologies, numerical prediction models, and expansive spotter networks to identify and warn on severe weather events. Many technical limitations inhibit the warning process as well, including radar beam overshooting and lack of upper air observations, to name a few. As was seen in Chap. 3 with public response, social forces also shape the decision making of institutional actors. Forecasting, emergency management, and the scientific weather enterprise do not take place in a psychological, social, or cultural vacuum. Nevertheless, while a great deal of empirical and theoretical attention has been given to public warning response, comparably less has been devoted to understanding how public officials—notably, emergency managers and NWS forecasters—make decisions regarding the warning process. Just as social, cultural, psychological factors come to bear upon warning response, so too do such forces influence decision-making processes of public safety institutions in both the short and long term weather contexts.

So acknowledged, in this chapter, we set out to examine how emergency managers, forecasters, and the media make warning decisions, calculate risks, and evaluate the weather, reviewing and outlining a wide range of factors at the institutional level that shape how actors arrive at judgements regarding risk and warnings. While advances in technology and bureaucratic sophistication have resolved countless forecasting uncertainties that in the past plagued decision making, such developments have also brought into view unanticipated problems, notably regarding the social and psychological dynamics at play in managing severe weather emergencies.

W. Donner et al., *Severe Weather Warnings: An Interdisciplinary Approach*,
https://doi.org/10.1007/978-3-031-05031-2_4

Indeed, overcoming many of the historical problems of meteorological detection and prediction have brought into light new uncertainties: How do forecasters think about the weather? What role does stress play in the decision-making process? How much data is too much to think about and manage? What is the best way in which to articulate risk information so that it might be best understood?

Obstacles to Successful Warning at the Institutional Level

The decision to warn is no simple affair. One source of this complexity is the sheer number institutional actors involved in severe weather detection, prediction, and warning response. Prior to warning communication, decision makers must first make sense of a myriad of sometimes conflicting weather information by relying on prior training and intuition aided by a broad range of data sources. These data and observations originate from, pass through, and circulate between countless stakeholder groups and individuals. NWS forecasters, local government officials, emergency managers, the media, and private weather companies each play critical roles in the larger "weather, water, climate enterprise". Each performs work in unique social settings and, as a result, are accordingly driven by different standards of practice, informal norms, values, beliefs, incentives, and goals. This amounts to the strong possibility that no two stakeholders will interpret and use weather data in the same way or for the same purposes. Stakeholder groups, for instance, often develop a unique set of heuristic rules or norms through which risk is judged. For example, Hoss and Fischbeck (2015) find that emergency managers favor visual cues over data as a means to confirm riverine flooding. Forecasters, too, through a combination of socialization and experience, are known to informally develop subjective forecasting methods through which to better manage the natural complexities and uncertainties that attend even the most robust weather data. What is more, different stakeholders possess distinctive interests that further govern the use of severe weather information. For instance, the business motives of national media institutions may compel private news companies to focus attention on larger, urban areas at the expense of smaller media markets. As early as the 1980s, and in part due to the increased politicization of the role, elected emergency mangers have increasingly tended to weigh the anticipated political liabilities that attend warnings (e.g., false alarms) against public safety (Petak 1985) Taken together, this shows that additional factors beyond mere data and technology shape the choices that institutions make regarding tornadoes, floods, and other hazards. There is always a context in which decisions take place.

Weather Forecasters

Official public warnings are issued by NWS forecasters. Weather forecasters are trained professionals paid to observe and interpret weather data for a wide range of threats, including, but not limited to, tornadoes, hurricanes, floods, and other forms of high-impact weather. Although forecasters train for a variety of different weather scenarios, here we focus mainly on those most relevant to the discussion of severe weather warnings: General and Lead NWS forecasters, both of whom are responsible for predicting and warning on severe weather through the analysis of weather observations. In simple terms, the main responsibility of forecasters is to monitor and predict atmospheric conditions, judging what future weather may hold by applying scientific training and intuition to the interpretation of weather observations and numerical model forecast output. In effect synthesizing the roles of both scientist and emergency manager, forecasters occupy a unique position within the organizational hierarchy of public safety, oftentimes serving as an ambassador of sorts between the scientific and practitioner communities. Many forecasters hold at the very least an MS in meteorology, and a good number of PhDs, largely due to the scientific professionalization of the field over the past century. Many forecasters in the private sector have BS degrees.

Before proceeding, it is important to note one key distinction in the forecasting community, specifically the distinction between public (i.e., NOAA) and commercial meteorological organizations (e.g., AccuWeather). In an historical analysis of private sector weather organizations, Spiegler (1996) notes that private weather organizations traditionally tended to provide businesses, utilities, and airlines—and other non-public institutions and entities—with information tailored to the needs of those organizations. For instance, airlines may need more frequent updates on wind information due to the critical impacts on takeoffs and landings. In contrast, the NWS holds a more general focus on public safety, and therefore concentrates on the broader purpose of issuing severe weather warnings to communities. Yet the distinction is at times vague and not easily defined. Increasingly, private companies provide warning information and products to news media, thereby taking on a broader role in public weather safety than seen in the past, even at times issuing severe weather warnings independently of government institutions (Smith 2002). Presently, no laws or policies regulate who may issue severe weather warnings, nor do governments at the local, state, or federal level regulate the form a warning may take. The absence of regulation grants a significant degree of freedom, but at the same time holds the potential to create confusion through the development of overlapping weather products. For instance, in 2019, the Sinclair Broadcast Group, a telecommunications conglomerate, suffered heavy criticism due to its use of what were termed "Code Red" warnings. Largely unsanctioned and with no relation to NWS warnings or watches, it was feared that the use of "Code Red" warnings would only serve to further confuse a public already somewhat unsure of the distinction between formal NWS alert types (https://www.cnn.com/2019/06/13/media/joe-crain-sinclair-code-red).

Few people outside of the weather community are familiar with what forecasters actually do. Before describing the literature on forecaster psychology and behavior, it is first necessary to provide a formal description of what forecasters do on a regular basis. The following is a description of a recent job posting for a forecaster posted by NOAA, offering a good description of common activities (https://www.commerce.gov/sites/default/files/2019-02/position-meteorologist-gen-forecaster-12.pdf):

1. During an assigned shift, the forecaster is responsible for integrating all meteorological data available from a variety of sources, and for analyzing and assessing the current and forecast weather situation at both the synoptic and meso-scale levels. The incumbent devises and formulates all necessary warning, advisory, and forecast products assigned to his/her forecast "desk" during the shift, with emphasis on meeting the needs of the users. . .prepares and issues warnings for the WFO service area (and coastal waters, if applicable) concerning a number of hazardous weather conditions, such as severe weather, high winds, flash floods, marine effects, winter storms, etc.
2. Conducts a weather watch which involves interpretation of Doppler radar data and satellite imagery, and the analysis of other meteorological and hydrologic data. Determines appropriate and relevant data, identifies questionable or conflicting data (resolving those if possible) and analyzes all integrated data sets to determine: (1) need of warnings, statements, or advisories, and (2) need for issuing or updating local public, and aviation (or marine, if applicable) forecasts. Maintains a radar surveillance of the service area using the NEXRAD system. Expands the use of NEXRAD data and products in the daily forecast and warning environment through the application of experience and locally developed studies and techniques. Fully integrates NEXRAD product and data advancements into the preparation of all WFO warning, forecast and public service products, including individual briefings.
3. As needed, provides weather advice and guidance to the general public, and to agricultural, commercial, transportation and recreational interests. Provides forecasts and weather briefings to press, radio, and television stations.
4. Provides hydrologic service products for assigned WFO area, including forecasts and warnings of floods and river stage for the public and numerous users dealing with water resources and/or land management, transportation, emergency management, river and flood plain control, etc.
5. Conducts or participates in the conduct of local studies and developmental projects designed to capitalize on or incorporate the benefits of new science/technology/local techniques into WFO operations. Applies this developing body of knowledge and experience to the warning and forecast environment.
6. Participates in fostering improved service through the establishment of relationships with users of special program forecasts or in Focal Point activities as assigned by the MIC.

This process changes dramatically when the potential for severe weather emerges. As far as severe weather goes, the process by which forecasters issue a tornado warning may be described in several distinct, analytic stages (Fig. 4.1 adapted from Brotzge and Donner 2013). First, a "tornado threat" is detected either by the NOAA Storm Prediction Center (SPC) and/or by a local NWS Weather Forecast Office (WFO). The role of the Storm Prediction Center (SPC) is to help guide the local WFOs in alerting them to the larger-scale, longer-term presence of a severe weather hazard threat. A threat of severe weather is often realized several days in advance, putting local regions on a higher state of alert. In general, the SPC provides daily guidance on where, when, and the likelihood of severe weather. Local officials

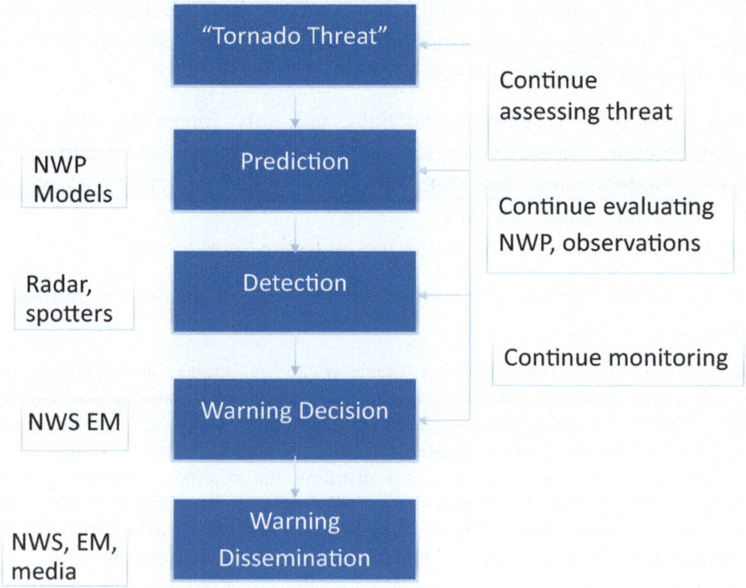

Fig. 4.1 Forecasting-to-warning process [adapted from Brotzge and Donner (2013)]

consider Storm Prediction Center (SPC) guidance in evaluating their local hazards. As a potential hazard draws near, officials shift from utilizing solely NWP guidance to inclusion of weather observations, from satellite, radar, and surface stations. If data and/or very short-range NWP suggest a developing hazard, a forecaster may shift to looking for certain patterns or features within the data to confirm the precise location and timing of the hazard; for example, a "hook echo" on radar is an excellent indicator of a developing tornado. At this point, a warning decision is made by the NWS and/or emergency managers, oftentimes in collaboration with local emergency managers, although emergency managers have autonomy to issue warnings independent of the NWS (it is often thought that emergency manager are required by law to issue warnings if the NWS issues a warning, but this is not the case).

What Affects Warning Judgements in General?

We now turn from a description of the formal forecaster role to a discussion of research on forecasters and decision makers—namely, the factors that govern how decisions are made within an operational setting. But why study forecasters at all? Computers themselves do not provide forecasts but merely "outputs [that] are specific predictions of certain weather parameters" (Hoffman et al. 2017: 333); it is left up to forecasters to decipher and interpret the suite of forecasts for any meteorological significance. In other words, the decision is not left entirely to technology. The technology provides information, but the forecaster must still summarize and interpret that information. As noted, the forecaster is an inseparable part of the warning process because

- Computers analyze meteorological data, but only forecasters can determine, based on context, whether given scenarios or features are significant;
- Computer models cannot integrate pattern anomalies or statistical outliers well, which forecasters adjust for; and
- In general, the forecaster adjusts the model to the "real world."

Doswell and Schultz (2006: 13) further reflect on the ineradicable human role in forecasting:

> In our experience, many severe weather forecasters and researchers are seeking a "magic bullet" when they offer yet another combined variable or index for consideration, whether or not they realize it. If some single variable or combination of variables made forecasting so simple, then the need for human forecasters effectively vanishes. There may be other reasons for the demise of human forecasters, but distilling the complex atmosphere with its nonlinear, possibly chaotic, interactions into an all-encompassing variable seems improbable. Any forecaster seeking to find such a variable is not only unlikely to be successful, but, if success were achieved, the need for a human forecasting that event vanishes!

Subjective sensemaking on the part of the forecaster is therefore a central task of forecasting practice. But what governs how decisions—this process of making sense of the weather—are made in the forecast office? Just as decades of research have shed light on public response (see Chap. 3), so too have scholars studied how forecasters survey and evaluate the weather—from how they collect and verify data, weigh model outcomes against personal intuition, and arrive at a verdict to warn or not to warn (although, it should be noted, that the literature on forecasters is comparably far more limited in scope, depth, and sophistication *viz-a-viz* public response, and as a result there is no analogous model of forecasting response as there exists for the general public).

In close accord with Kahneman and Tversky's classic theory of risk psychology, research shows that heuristics—informal rules or "rules of thumb"—factor to varying degrees into institutional decision making with respect to severe weather. In the past, forecasters relied heavily on heuristic rules, developed through experience and training, to recognize and correct for the ambiguity and noted imperfections associated with modeling process (Roebber and Bosart 1996). This is because no model, however useful, returns a perfect representation or prediction of the weather. These heuristic techniques developed over time, sometimes through a process of scientific trial and error, and always within rules and confines of human psychology, culture, and social environment. However, the use of heuristics, while a feature of the past, is evolving among professional forecasters in more recent times. Today, while heuristics does play a part in the decision making process among forecasters, increasingly all offices share the same techniques and rules; it is becoming less common to use these 'old style' techniques as interpretations become increasingly a standardized affair. Technology is fast replacing these old rules. On the other hand, among emergency managers, the role of heuristics is far more common and pronounced.

Hoss and Fischbeck (2015: 90) describe an example of a heuristic technique used by one emergency manager to evaluate whether to warn on a flash flood: "When the foam is on the middle of the river, it is coming up. . .If I see big trees coming down, I know it is a lot water. If I don't see big trees, well, it is not a lot [of] water." Here the emergency manager in question, at the time when a decision must be made, relies heuristically on past experience and knowledge as a final arbiter in the decision making process. The data, while a useful guide, particularly during the early stages of an event, is ultimately of limited use when it comes down to the final decision among emergency managers to issue a local alert. Thus, heuristics, or, once again informal rules that guide the assessment of risks, exist because they are a functionally necessary feature of hazard prediction forecasting. With this established, the question then becomes: *how* do informal forecasting rules come about? To address the matter here we explore the intersection between psychological heuristics and the dynamic social and cultural settings in which forecasters make predictions and evaluate extreme weather scenarios under circumstances in which the need for rapid decisions under conditions of uncertainty is a normal feature.[1]

First, there is some use in describing how weather forecasters think about and solve problems. Scientists from different disciplines approach thinking about questions central to their fields in vastly different ways. Within some fields, especially those focused on applied prediction, the employment of "mental models" is common. Sometimes referred to as *subjective forecasting* in the literature, mental modeling draws on heuristic thinking to work through forecasting scenarios naturally attended by high levels of uncertainty and ambiguity. In what is arguably the most exhaustive and detailed treatment of forecaster psychology, Hoffman et al. (2017: 241–242) argue that weather forecasters routinely must "envision a dynamic mental model of the weather" through an evolving process of "inspecting and integrating a great number of complex visualizations and data sources." This is to suggest that weather forecasters are given to contemplate the central problem of severe weather and weather in general in a manner fundamentally different from other forecasting scenarios, such as, for instance, financial market forecasters, who may rely more heavily on numbers to guide prediction. Roe (1951), studying how scientists conceptualize and imagine problems, finds that biologists and experimental physicists tend to think problems through with the aid of mental imagery and models, whereas physicists, psychologists, and anthropologists normally conceptualized ideas in verbal terms.

How, then, do weather forecasters specifically think about tornadoes, floods, and other forms of severe weather? In other words, how do they conceptualize or "think through" decisions regarding the weather? According to most scholarship, weather

[1]While heuristics up until now has played a significant role in forecaster judgement, it should be noted that the growth and increasing application of artificial intelligence in forecast modeling may, in the future, fundamentally alter the role of the professional forecaster. How, and in what ways, remains to be seen.

forecasters likely characterize and fathom the weather in graphic or visual terms, but in a way far more complex and far more dynamic than simply conjuring mental images of weather patterns and structures. In this regard, Miller (1972: 11) suggests that the most successful weather forecasters tend to "carefully analyze, coordinate, and assess the relative values of a multitude of meteorological variables, and mentally integrate and project these variables three-dimensionally in space and time." By way of comparison, whereas financial forecasters rely heavily and perhaps almost exclusively on numbers and graphs to make market predictions, weather forecasters, while also relying heavily on quantitative reasoning, assemble from data an overall, evolving mental image of weather conditions. Hoffman (2017: 240) provides what is perhaps the best summary of how forecasters conceptualize and predict the weather: "[weather forecasting involves] combining different perspectives, modifying a mental image by adding or deleting features, mentally moving or rotating an object (air mass, front, etc.), and mentally animating static images and projecting into the future." From an interdisciplinary perspective, that forecasters apply visualization techniques is critically important since technologies should be adapted to this unique way of working through forecast uncertainty with respect to the weather. In short, the basis of meteorological forecasting involves building visual, mental models through a synthesis of forecaster intuition and meteorological observations.

How then are meteorological data (e.g., numerical weather prediction, rainfall estimates, windspeed data) integrated with the heuristic, mental modeling process to provide an overall assessment of the weather? Indeed, in the practice of weather forecasting, considerable interplay occurs between subjective forecaster assessment and meteorological data gathered from instrumentation—oftentimes when there is too little or two much data. Through an analysis of Quantitative Precipitation Forecasts (QPF), for instance, Funk (1991) finds that objective forecasts—those based on numerical predictions and estimates—are often tempered by "rules of thumb," or what we refer to here as subjective forecasts. These rules of thumb are found to be applied contextually, not uncommonly when formal meteorological models yield incomplete or vague descriptions of regional weather patterns. By way of example, Funk notes that "If heavy rainfall is expected during periods of weak synoptic and/or mesoscale forcing (e.g. many summertime events), then the latter techniques (rules of thumb) may become increasingly important" (563). In other words, data that defy the expectations of circumstances often call for the heavy use of heuristics—or "rules of thumb"—on the part of forecasters. Data indicating the possibility of tornadic storms outside of tornado season, for instance, may be more heavily scrutinized using subjective forecasting.

While often used to resolve gaps in meteorological data (e.g., radar information blocked by a building or mountain), subjective forecasting often arises in the context of *too much* data, as well. Stuart et al. (2007: 1894) observes that "forecasters are faced with so much information from observations and from multiple numerical model forecasts that in some circumstances an analytical decision-making process is limited by time constraints. Thus, heuristic approaches are virtually mandatory for weather forecasters." Doswell (2004: 1117–1118) concurs, also noting that the routine employment of subjective forecasting may in part be due to an information overload of sorts:

> Given modern observation and analysis technology, human forecasters have compared the flood of new in-formation in forecast offices to trying to drink from a fire hose; the torrent of data and products derived from the data threatens to overwhelm them. . .Forecasters have become concerned about knowing what are the most pertinent and helpful products to examine and use, and just precisely how to use those products when forecasting is subject to operational forecasting deadlines. . .but whenever there is no known objective path to the optimal use of information, human weather forecasters must rely on their experience and intuition derived therefrom, that is, heuristics.

This "flood" of meteorological information is of concern since the gross quantity of data available at any given time remains only as useful as forecaster skill and human limitation allows. By way of example, higher resolution radar data may, in theory, offer a more accurate picture of the weather, but the ability to interpret meteorological patterns within higher resolution remote sensing data is known to be deeply constrained by forecaster skills *per se* (Roebber et al. 2004). Thus, in a practical forecasting context, resolution beyond a certain level of detail may yield diminishing returns on the quality of prediction in practice. This is not to say better resolution is undesirable, but rather to acknowledge a potential "ceiling" effect imposed on interpretation quality that may be a function of forecaster training or human limitation. In addition to improvements in data resolution, questions have also emerged regarding the rate at which data arrive in the hands of forecasters—updates to radar data as storms evolve being of particularly interest in this respect. Namely, after an area of interest is scanned, and data are returned, radars must go through a full rotation until that area is once again scanned. If our hypothetical radar requires 5 min to fully rotate, forecasters will not be able to "see" new data for the area of interest for an additional 5 min. While it may seem natural to reduce update time, questions have arisen as to whether doing so would give way to a psychological tendency on the part of forecasters to delay decisions in anticipation of another updating being right around the corner, as it were. Heinselman et al. (2015: 74) reflects: "it is possible that 1-min updates could lead some forecasters to postpone their warning decisions to wait for 'just one more scan.'" Longer updates, on the other hand, although providing less information, force action, albeit with less-up-to-date information on the storm system of interest to the forecaster (hence the need for heuristic rules in this scenario). From an interdisciplinary perspective, a key question would be to determine the ideal update time that would balance well the need for rapid decision against shorter update times in order to aid the process of mental modeling intrinsic to the forecaster thought process.

In addition to a rapid growth in the quantity and volume of data to which forecasters are exposed, the quality of data meteorologists encounter also affects the development of mental models of severe weather. One pressing question in the literature relates to whether deterministic forecasts or numerical uncertainty (i.e., probabilistic) forecasts assist operational forecasters in making better judgments regarding the likelihood of events. Joslyn and Grounds (2015) find that numerical models in which uncertainty is expressed in probabilistic terms yields better decisions in comparison to deterministic forecasts. Accordingly, they find that in a series of experiments in which subjects were tasked with deciding whether to salt roads based on forecasts, it was found that those subjects who encountered uncertainty information made better decisions than those presented with deterministic information. The authors suggest that this is due to a fundamental distrust of deterministic forecasts. Also observed was a tendency among subjects to impose their own rulings on the probability of events when confronted with circumstances regarded as vague or uncertain, suggesting that individuals likely think in probabilistic rather than deterministic ways about the weather. Results similar to these were observed in earlier research by Savelli and Joslyn (2013).

Other research focuses on how forecasters and meteorologists develop aptitudes over time. Scholarship suggests this occurs through a combination of formal learning and socialization. In this regard, Bond and Mass (2009) examined the process by which senior meteorology students acquired and developed forecast skills over time. According to findings, the average student required roughly 6 weeks to develop proficiency in forecasting next-day atmospheric phenomena (clouds, rain, temps, etc.). Beyond this 6-week period, however, data suggest only marginal growth of forecast skill, a plateauing effect similarly discovered in Suess et al. (2013), though here much more rapidly (roughly 2 weeks as opposed to 6 weeks). Overall, the greatest gains in proficiency seemingly resulted from the routine drilling of meteorological examples, but also, importantly, through gaining an understanding of the context of local weather through interaction with local forecasters through the process of local socialization. Thus, while forecast skill is heavily tied to training and coursework, talents were also cultivated through a process of interacting with local meteorological staff. Indeed, this is perhaps not surprising, as experience, rather than training alone, has been found to be one of the strongest predictors of forecast skill (Roebber and Bosart 1996). In light of earlier discussion regarding heuristics, these findings appear sensible: it is through experience rather than formal education that the informal rules that shape the interpretation of meteorological data are acquired.

As a final note, looking to the future, as meteorology moves forward technologically and sees an increasing number forecasting operations automated, questions about the role of humans in the forecasting process have been raised. Novak et al. (2008) conducted a survey of 237 operational forecast managers in a variety of forecast settings, including National Centers for Environmental Prediction's (NCEP) National Centers (NCs), Weather Forecast Offices (WFOs), and River Forecast Centers(RFCs). The goal of the survey was to understand the changing role of the weather forecaster in an increasingly computer-dominated profession. While most respondents saw a clear role for forecasters in this new environment, there was significant disagreement over the parameters of this role. Whatever the future holds for computer-driven meteorology, there was consensus, however, on the implicit need for continued involvement of human forecasters in communicating and evaluating forecast uncertainties, interpreting forecasts, discussing alternative forecasts, deciphering probabilities, and ultimately tailoring products to user needs. Computers, automated algorithms, and AI, while useful from a technical standpoint, cannot directly interpret or make sense of the weather for the public it was generally thought. There is, however, no consensus on how these forecasts should factor into the decision-making process from an operational standpoint. Half of those surveyed believed that the forecaster should take an active role in shaping automated system guidance, whereas the other half believed that the forecaster should only be involved in communicating the uncertainty of these interpretations. Novak et al. notes that much of this came down to how the individual forecaster perceived the system.

Emergency Managers

Emergency managers are responsible for a wide range of activities, from preparedness, mitigation, response, and short- and long-term recovery. Critically important among these tasks, especially among communities that routinely face the risk of severe weather, are the development and administration of protocols to mitigate the impacts of tornadoes, floods, and severe thunderstorms. When severe weather strikes, or is thought to be immanent, emergency managers monitor the weather, coordinate with NWS forecasters, and issue local warnings and alerts if necessary. In accordance with U.S. federalist doctrine, although guidance is often sought through meteorologists and forecasters, the ultimate decision to warn at the local level rests with the local emergency managers. There is currently no law at the federal level or elsewhere that requires an emergency manager to issue a tornado warning should a warning be issued by the NWS. In most cases, in a local context, warnings are issued in the form of siren systems (League et al. 2010). There is no single type of emergency manager per se. Some emergency managers have access to well-supplied and well-resourced EOCs, modern equipment, and personnel. Other emergency managers may have access to far more humble resources, conducting operations on a volunteer basis on a personal computer. As Hoss and Fischbeck (2015: 87) note, most emergency managers hold backgrounds in fire and police work, with considerable variation in safety resources, with one emergency manager in the study enjoying access to a "fortified office, staff, jeeps, trailers, and boats" while another emergency manager had at their disposal simply a "phone."

When it comes to the decision to warn communities, for the following discussion it is important to note that local emergency managers are not bound by any law or regulation to issue a warning (largely due to the federalist doctrine that "all response is local," which places the burden of first-line response mainly on local emergency managers). As a result, there is wide range of criteria that emergency managers use to decide on whether to issue a warning. Some emergency managers may issue warnings if local NWS forecasters have done the same, others rely on verification through local spotter networks to verify tornadoes, while still others monitor the media and warn accordingly. In effect, the decision to warn or not to warn is made on a case by case basis, taking into account a wide range of factors—some static, others dynamic—into the decision making process. Here we explore some factors that shape this decision-making process.

To Warn or Not to Warn: Decision-Making Patterns Among Emergency Managers

As far as decision making goes, much like forecasters, emergency managers also appear to rely on heuristic tools and informally-developed norms that emerge through a combination of intuition and training. The rules that emergency managers use to decide whether to warn or not to warn are far from simple, however, oftentimes taking into account not merely weather conditions, but also political and social contexts in what is in reality a process that is irreducible to any one single factor. For instance, when asked whether they would prefer more refined tornado or rainfall estimates in new systems (as well as justification for the choice), emergency managers appear to take into account a wide variety of situations, circumstances, and contexts, and while the inherent danger of tornadoes often weighs heavily as a defense for better tornado data, it was not the sole factor that drove stated preferences:

> First, it is the perspective of emergency managers that flood and rainfall threats imperil communities less these days due to effective mitigation. While more resistant structures mitigate tornado impacts well, there are a broader and more effective range of flood-protection options available to communities. Second, greater levels of uncertainty are associated with where tornados will occur, where they will end, and how long they will last—accurate and effective tornado tracking has yet to be developed and implemented. On the other hand, as one emergency manager cogently states, "[F]loods are more of a nuisance than anything." Tornados, according to a majority of emergency managers, should have more resources devoted to them because of their irregularity, while flooding can be predicted relatively well. One emergency manager comments on this issue: "Whether I have the other resources or not I'm still going to concentrate on it [the tornado] because that's the most unpredictable catastrophic of the two events, even though given the right conditions the flood could be more catastrophic, history tells me here that it's not." Another emergency manager expresses his views on allocating resources: "[W]e can monitor the flooding threat or the rainfall threat pretty well. And so I would be biased towards allocating, splitting the resources and putting more resources into the tornado detection." Finally, more accurate tornado information would enhance the credibility of emergency managers because the public relies heavily on tornado warnings in order to make personal safety decisions in times of severe weather. Thus, there appears to be an underlying political motivation behind preferences for tornado detection over QPE. One emergency manager indicates that his job depends on the warnings he is able to provide: [I]f I don't give the warning for a flood, I'm still going to be here tomorrow; if I don't give the warning, if I don't blow the sirens before the tornado hits the city limits, I won't be here tomorrow (Donner et al. 2005: 6–7).

Indeed, as far as warnings are concerned, emergency managers appear acutely mindful of the potential political consequences and liabilities brought on through repeated errors in judgment. Indeed, concerns over credibility may lead some emergency managers to use forecast types known to be less effective in the warning process. Demeritt et al. (2010) notes that while deterministic forecasts (e.g., "the event *will* happen") are less helpful relative to probabilistic forecasts (e.g., "there is a 60% chance an event will happen") in an operational context, emergency mangers and local forecast users nevertheless held a preference for deterministic forecasts because the use of which produced less accountability under the circumstances of missed events or false alarms. Especially for emergency managers, deterministic forecasts may displace accountability back to the originating forecast office, thus reducing the blame immediate decision makers would otherwise bear, and thus reducing political costs, for false alarms or missed events. On the other hand, probabilistic forecasts require interpretation and judgement prior to issuing a warning and therefore shifts accountability back to the emergency manager. As a result, if the warning turns out to be false, or an event is missed, greater costs to credibility will be suffered. With deterministic forecasts, emergency managers are for all purposes less vulnerable to such risks.

With more sophisticated communication technologies, emergency managers enjoy unprecedented access to remote sensing data in both raw and product form. Emergency managers presently use radar data and radar data products for a wide variety of activities, including directing spotters and search and rescue (Kuster et al. 2017; League et al. 2010), as well as evaluating the decision to warn the communities over which they hold jurisdiction. Emergency managers, however, do not seem to prefer raw, uninterpreted radar data, but rather what are known as "radar products" that simplify and translate the radar data into easily digestible, quick-updating visualizations. This may be because many municipal emergency managers in particular lack the training and expertise to make use of raw radar data, and therefore require the original data to be rendered in a format more immediately comprehensible among users with less training and meteorological expertise. This notion is illustrated in Baumgart et al. (2008), who observed that emergency managers made heavy use of weather radar reflectivity and velocity data, but only if experienced and well trained. It should also be noted that even with the growing use of radar data, emergency managers still appear to rely heavily on "ground truth" in the form of spotter observations (League et al. 2010). Indeed, these findings are reflected more recently in Weaver et al. (2014: 202), who observe a general hesitation among emergency managers to use radar data as a single and exclusive basis for warning decision making:

> Of the 883 managers who completed this section, 30.3% said they would activate their entire system with a Doppler-confirmed tornado a little over 30 minutes away. Receiving confirmation from law enforcement brought the statistic up to 57.0%, and when the tornado was roughly 20 minutes away and actually shown on television, the number jumped to 77.8%. The most frequent reason that the remaining 22.2% gave (in a block provided for verbal clarification) for not activating full response was that they would not fully engage until damage had actually occurred in their area (53 such specific answers). This reticence may have to do with survey results that clearly show a large majority of emergency managers to be focused by education, training, and experience to be responders.

Central to the warning process, factors regulating the flow of information within and between emergency management institutions has also been touched upon in the literature. Radar and warning information may be of the highest quality, but if degraded in accuracy and reliability through the process of communication, will be of little use. With these concerns in mind, Hoss and Fischbeck (2015) interviewed 45 emergency managers in Pennsylvania, Oklahoma, and Arkansas on issues related to riverine risks. Results of the study demonstrate the oftentimes complex interrelationships between NWS, local emergency management, and local city government in the exchange of information, knowledge, and observations within and between key stakeholder institutions. In this regard, an important distinction between the NWS and local emergency management is noted, particularly with respect organizational complexity. Insofar as NWS is concerned, there are generally fewer members through whom information must circulate and by whom certification is necessary. Thus, all else equal, weather information, while nevertheless going through a process of interpretation and analysis on the part of forecasters internally, tends to enter and exit NWS institutions *relatively* undistorted. On the other hand, even local government, the context within which local emergency managers work, suffers from a great deal of political complexity—comprised of diverse number of roles, groups, and institutions with an equally diverse set of norms, values, and interests. During an emergency, emergency managers must often pass information through this complex organizational structure, a process that may alter key facts and interpretations about the hazard at hand. Even the most well-worded and clear warning messages likely undergoes corruption of meaning to some degree when the channels through which it circulates expands in number, character, and complexity.

A final note on funding should be made. What must also be understood about emergency management in the United States is that under federalist doctrine "all response is local," meaning that local government bears the responsibility for managing disasters, warning response included. Although federal agencies such as NWS provide information, such as forecast and radar data, much of the assistance is indirect; ultimately decision making comes down to local officials. From an interdisciplinary perspective, it is important to note that emergency managers and the EOCs they run are as diverse as the communities they serve. Hollywood movies such as *Volcano* often portrays emergency managers as professionals with access to technologically advanced EOCs, which are adorned with multiple high-end computers, flat screen TVs, and staffed by a trained collection of personnel. While this might accurately characterize some emergency management EOCs, particularly in affluent or urban settings, a vast majority of local EOCs operate on shoestring budgets and are manned by fire, police, or volunteers. Thus, one-size-fits all integrative approach will fail under such circumstances.

Media

The media plays no small or insignificant role in communicating risk during periods of severe weather. Quarantelli notes that communicating warnings is the principle role of media institutions during disaster (1991). Despite a growth of social media, studies still suggest that most people will turn to television and radio for disaster-themed information (Burger et al. 2013), although recent research on the issue may prove otherwise given the rapid expansion of social media as an information tool in the past decade (see Chap. 8).

In comparison to forecast institutions and emergency managers, relatively less work has been done on media institutions and their role in warning response. The media is a unique stakeholder in the overall schemata of warnings, for, unlike emergency managers and forecasters, media institutions technically operate without a formally sanctioned role or mandate within the warning process. For analytic purposes, one might compare the media to emergency managers given that both issue warnings as a matter of practice, yet media institutions are not civil servants as are emergency mangers and therefore remain largely autonomous in the warning process. (Unlike emergency managers, commercial weathermen cannot be "elected out" or removed from office due to poor performance, so there are no political checks and balances in this regard). Media institutions may, for instance, communicate National Weather Service warnings to the public and do so without approval from local, county, or regional emergency management. For instance, a local news station may interrupt a television program to communicate an NWS warning or, during a regularly scheduled forecast, discuss the potential for severe weather. As such, the media serves as an important channel for communication, although not one without problems and issues. It should be noted that news meteorologists typically do not create their own forecast products from raw meteorological data, but rather communicate pre-existing warnings to the public. Demuth et al. (2012: 1137) summarize nicely the general role of the media during severe weather:

> [local news meteorologists] and radio broadcasters rely on their warning system partners—NWS forecasters and emergency managers—to provide them with hurricane risk information. The media also rely on private sector weather vendors to transmit and postprocess NWS data, produce value-added forecast information, and provide a platform for creating camera-ready graphics for television. The media say they then take the hurricane risk information they receive and focus on "boil[ing] it down into bullet points, into facts, into useful information" for their audiences. As one television meteorologist described, Most people today . . . still get their information from TV and still rely on us to explain it to them, decipher the information, and communicate it.

The media thus act as sort of a value-added conduit or, in Demuth et al.'s (2012: 1136) regard, "storytellers" serving as translators of NWS information for a public audience. At the same time, media institutions, unlike emergency managers, are for-profit businesses and hold no political allegiance to any one jurisdiction, and therefore may focus attention on larger, urban media markets at the expense of smaller, rural markets when the possibility of severe weather arises (Johnson 2020). Indeed, Wenger (1985) was perhaps the first scholar to examine the relationship between the media and disasters, noting the at-times contentious, other-times cooperative "friend-or-foe" relationship between the media and politically-endorsed institutions involved in the warning process.

It is also known that the public forms strong attachments to certain media figures, and weathercasters are no exception to this rule. Because media personalities tend to be more visible and selected for a combination of appearance and charisma, this has resulted increasingly in what scholars have termed the growth of "parasocial interaction" between media forecasters and audiences (Sherman-Morris 2005). Horton and Wohl (1956) describe this form of simulated intimacy between viewer and weathercaster as being akin to a "face-to-face relationship between spectator and performer," which develop into bonds of trust between viewers and weather forecasters over time. Moreover, such bonds in turn go on to cultivate trust in media figures with whom sustained parasocial "interaction" has taken place, which may conceivably further affect behavior in the form protective action, although, it must be mentioned, the role of parasocial connections in motivating warning response behavior remains presenlty in dispute (Sherman-Morris et al. 2020). Nevertheless, the risk under such circumstances is that weather forecasters, due to the formation of social bonds, may be taken more seriously by the public in comparison to public officials, who are notably less visible and perhaps less dramaturgically-compelling than the average weathercaster, who tend to be selected for these traits.

What Governs the Warning Process at the Institutional Level?

While the previous sections dealt with the specific circumstances surrounding the decision making of forecasters and emergency managers, this section discusses the more general features that govern severe weather decision making whether one is an operational forecaster or emergency manager.

Psychological

Individual psychological factors play a natural role in the management of disasters. It should be of perhaps little revelation that emergency managers and forecasters, especially during acute periods of severe weather, face tremendous stress and anxiety. The consequence of issuing a false alarm—or, worse, missing an event entirely—holds profound consequences as far as public credibility and safety are concerned, not to speak of the moral mandate of saving lives. Quite literally a matter of life and death, the ongoing assessment of severe weather and decision to warn presents a source of growing pressure in the days and hours leading up to a severe weather outbreak. Indeed, stress is likely due to the reality that meteorological decision making is inherently "characterized by couplings between alternatives and outcomes that are probabilistic and thereby cannot be predicted with certainty" (Maule and Svenson 1993: 9). How do factors such as stress, anxiety, and pressure alter the decision-making process of emergency managers and forecasters? How does it render the ability to detect, predict, and warn of hazards and risks? This has been a matter of debate for several decades.

One of the earliest attempts to study the role of psychological pressure in the decision-making process among responders comes from Wright (1974), who found that decision makers, when stressed, increasingly favor negative information and rely on fewer sources of information in the decision making process. Stress was found to create what is commonly referred to in lay terms as "tunnel vision," or a narrowing of focus on certain events or sources of information at the expense of the "big picture." Indeed, similar results are summarized in Rothstein (1986), as well as Edland and Svenson, who observe that "studies of judgement under time pressure seem to indicate greater weight given to less positive and more important attributes, greater variability of judgements, and a tendency of using a smaller number of attributes" (1993: 29). Furthermore, the major risk when it comes to forecasting appears to be an overreliance on heuristics or behavioral routines in the evaluation of meteorological data and the risks implied therefrom. This is important here because while heuristics and informal rules, as noted earlier in this chapter, serve an important function in the forecasting process, they may be overused during periods of intense psychological pressure. Accordingly, Betsch et al. (1999) found that in an experimental study that subjects, despite exposure to new, contradictory evidence that might justifiably alter decision making, would oftentimes rather default to common rules as a guide to judgement when under circumstances where the need for rapid decision making was imposed. Under conditions of stress, established norms offer a "way out" of the building anxiety and therefore hold greater allure in the decision making process in tense decision contexts. This may lead emergency managers and forecasters to default to rules rather than new information when pressed to make rapid choices.

Experiments involving simulated emergencies shed some additional light on the matter. Dorner and Pfeifer (1993) developed a series of simulated games that required two groups of subjects to control fire outbreaks by directing hypothetical fire personnel to patrol, search for fires, or extinguish fires. In order to imitate stress experienced under such circumstances, experimenters exposed one group to white noise and sounds of varying length and intensity throughout the exercise, whereas the "control" group was not exposed to these conditions. In general, both groups saw equal success when controlling the fire: results found no difference between stressed and non-stressed groups in performance. However, each group developed different problem-solving approaches to managing the simulated crises. As discussed in the previous paragraph, and perhaps most notably, the non-stressed group focused on general, "big picture" problems, resisting the tendency to linger on any individual outcome or piece of information for too long. The stressed cohort, on the other hand, adopted a more narrow and focused approach, devoting much more attention to individual units and less to the general situation. Despite dramatically different management approaches, it is surprising that both groups responded equally well to the simulated fires, though this may be because the simulated activity was fairly simple, undemanding, and obviously known to be contrived on the part of participants. Detecting, predicting, and warning for severe weather, which requires general attention to numerous kinds of data, an ability to integrate a surplus of weather observations, and the coordination of multiple stakeholders, on the other hand, likely demands "big picture" thinking due to the presence of so many "moving parts," as it were. A forecaster therefore cannot become preoccupied with any single piece of information or event for too long. In the context of severe weather prediction, no single piece of information is useful in and of itself, so the focusing effect of stress may have a more negative outcome for decision making in the context of forecasting and emergency management.

Social Factors

Involving interactions between individuals, groups, and institutions, the decision to warn is not merely psychological in nature, but also of a sociological character, going beyond issues of stress, emotion, and cognition. Social bonds, for instance, which represent attachment commitment, and involvement between groups (so defined by the famous sociologist, Hirschi), play a key role in the transfer of information, trust, and decision making between institutions, of which local emergency management and NWS are no exception. Ernst et al. (2018), in a recent study, sought to understand the use of weather information among a sample of 11 emergency managers. Unsurprisingly, social bonds featured prominently in the analysis. But in what ways? First, it was observed that emergency managers relied heavily on local NWS forecasters for weather information, especially as severe weather drew nearer and even more so when there was uncertainty present in the weather data. Secondly, far from being merely instrumental, these relationships were, as perhaps the term heavily implies, social in nature, cultivated over time through interaction and personal association. Importantly, the exchanges that were ranked as most useful among emergency managers were those that took the form of live interaction with NWS personnel. Over time, the bonds that arose with local forecasters brought forth greater trust and confidence among emergency managers in regional weather forecasts, predictions, and data, information that may have otherwise been regarded with skepticism under circumstances in which forecasters remained indifferent or aloof to the needs of local decision makers. Demuth et al. (2012: 1137–38, current authors' emphasis) report similar findings, noting that "The emergency managers we interviewed spoke about their strong relationships with specific NWS and media partners, including personal connections that helped them contact WFO forecasters. . .the partnerships are not flawless. . .nevertheless, an important component of the partnerships' successes is the confidence that members of the groups have in each other based on the *organizational trust and personal relationships* they have built over time." Here we again return to the idea that the weather is as much about society and social relationships as it is about technology and the environment.

Indeed, the partnerships and social relationships between emergency management and NWS have been shown to be critical to successful warnings. Uccellini (1995) observes that during the 12–14 March Superstorm, warnings arrived with unprecedented lead times, which the authors attribute to coordination between NWS and emergency management at the local and national levels. Further, during a simulated experiment in which emergency managers were tasked with using radar to assess a tornado, many emergency managers called NWS to help interpret radar velocity data (Baumgart et al. 2008). Strong relationships between NWS, forecasters, and local emergency managers remain a cornerstone of success.

Cultural Obstacles

Finally, there are several cultural factors at play in the management of severe weather. The classic definitions of culture in sociology comprises *cultural norms, cultural values, and cultural beliefs*. Culture was earlier found to operate at the level of public response. In Chap. 3, for instance, a cultural belief in tornado mythology was documented in the literature, and more generally one does not have to look far beyond the more recent disasters for evidence suggesting that myths are alive and well today, attesting to the enduring impact of culture on risk perception and warning response. Following Hurricane Katrina, for instance, Tierney et al. (2006: 73) noted that the "lawlessness and social breakdown" narrative remains a well-established, though ultimately false, cultural attitude in society's collective understanding of disaster. The question here, however, is whether *emergency institutions* also hold these beliefs or whether they are largely immune to disaster myths.

Research suggests that some emergency managers do indeed hold belief in disaster mythology, though factors such as professionalization and experience play a role in determining the degree to which this is the case. Asking a group of emergency manager about expectations regarding public response during decontamination procedures, Carter et al. (2014) report that while inexperienced emergency managers anticipated mass panic following a decontamination response, more experienced emergency managers expected a more rational public reaction. Fischer (1998) encountered similar findings, noting that among all the myths that emergency managers hold, it appeared that panic and irrational behavior ranked among the most widely held. As recently as 2013, "mass panic" was thought a possible reaction among the public within a sample of public safety officials in Drury et al. (2013), although the degree to which such beliefs inform policy and decision making is largely unknown. The problem with the present literature is that many of the studies deal primarily with the immediate aftermath of disaster rather than the warning period itself. As a result, and while there is some generalizability to the results, it is unclear to what degree emergency managers hold disaster myths and to what degree they factor in the warning decision process. Further research is needed in this regard.

Differing organizational cultures may lead representatives from different institutions to "speak past one another." With this in mind, another key problem lies with the fact that meteorology is a scientific institution and all institutions possess unique cultures within which specialized terminologies develop, evolve, and come to define the interactions between members. The problem is that expert language, while a useful tool *within* an organization to efficiently communicate ideas, may be incomprehensible to those outside of the institution, and therefore requires significant translation and explanation so as to hold meaning for the general public. Earlier, Rogell (1972) noted that the present meteorological and forecasting vernacular may by-and-large be misunderstood by the public. More recently, surveying emergency managers, NHC, WFO, and media personnel in Florida, Demuth et al. (2012: 1139) summarizes:

> One reason this occurs is that NWS forecasters' primary expertise and training is as physical scientists. Moreover, they work in an organization that focuses on precise scientific information and that evaluates forecasters' performance based on specific metrics that are often scientifically based. As a result, they do not want to omit important details about hydrometeorological threats, and they sometimes have difficulty explaining those threats in layperson's terms.

Adding further weight to this perspective, Goldsmith et al. (2012) note that great success was made with plain language used during a rare winter event in South Texas. Warning response must pay careful to the linguistic norms of different institutions and non-expert communities.

Beliefs about the influence of repeated false alarms on public psychology also seem to influence emergency manager and forecaster decision making. What is important to note here is that whether false alarms have an actual effect on the public is irrelevant (which they do, see Chap. 3), when emergency managers act as if they do, this affects their use of technology. Several years ago, Demeritt et al. (2007) looked at the relationship between probability-based ensemble forecasting (EPS) and older deterministic forecasting for flood warnings. Even though the EPS warnings were found to be superior from the perspective of applied forecasting, as are probabilistic forecasts in general (Ramos et al. 2013), emergency personnel were hesitant to use the more reliable EPS models as a basis for a final decision in a simulated weather-scenario environment. Contrary to expectations, emergency personnel tended to rely on existing deterministic models as a final decision-making tool, using the EPS models not in their own right but rather to confirm the inferior deterministic models. As Demeritt et al. (2007: 126) explain, these observed preferences may come down to culture:

> In part this response reflected concerns about the costs of crying wolf'. In addition to the perhaps understandable desire to protect their own institutional reputation and credibility, the reluctance of flood forecasters to issue early warnings also reflected important beliefs about the cognitive and other capacities of the groups to whom they issue flood warnings. During the forecasting exercises as well as in the subsequent debriefing discussions and the questionnaire surveys, flood forecasters repeatedly ex-pressed the view that general publics, and even emergency service organizations, would not tolerate false alarms.

McCarthy et al. (2007) observed similar findings in their flood risk communication experiment, finding that most emergency managers felt uncomfortable with probability information. Thus, while probability forecasts may be better than deterministic forecasts, cultural barriers to their use remain in place.

Conclusion

Explored in this chapter were the institutional elements of warning response. Discussed were the unique ways in which forecasters think about the weather, the merits of weather information in the decision-making process, and the various psychological, social, and cultural elements that shape and potentially shape the warning process. From this, what can we say in the way of recommendations? We might distill from the literature several general practical policy recommendations:

1. *Focus on visualization or data that aid in visualization*: Although great strides in the visualization of weather data have been made, the primary focus of development should be in the creation and refinement of technologies that directly assist forecasters in building and articulating mental models. As noted earlier, different fields conceive of problems in different ways. Some inquire using mental models and imagery, others with numbers, and still others regard problems in verbal terms. Given that meteorologists have been shown to use mental models, it should perhaps not be surprising that forecasters articulate the weather mentally in imaginary and intuitive ways. Indeed, mental models oftentimes supplant objective models (e.g., numerical weather prediction) given regional cultural norms or if formal models are thought to be inadequate to context. Perhaps one day numerical models will achieve a level of sophistication necessary to eliminate the need for forecasters, thereby making the need for mental models moot. However, given the complexity of the weather, that day is long off. Therefore, at present, an emphasis should be placed on providing the tools forecasters need to develop mental and physical visualizations of severe weather.
2. *Assess the use and limitations of probabilistic forecasts*: The consensus in the literature appears to suggest that for both emergency mangers and the general public, probabilistic information is more useful in communicating an understandable level of risk, and, accordingly, leads to better decision making (though for balance, it should be noted that Demeritt et al. (2014) suggested some disagreement in the literature over this claim). Forecasters and emergency managers should make an effort to release probabilistic forecasts to the public and explore further strategies for clearly communicating risk probabilities to a public likely unfamiliar with the concept of probabilities.
3. *Examine more closely the relationship between updates and decision making*: Faster updates may be a double-edged sword. On the one hand, fasters updates allow for a better assessment of the present weather situation, which, in turn, allows for more accurate warnings, a reduction in the False Alarm Rate (FAR), and fewer missed events. On the other hand, there is the general concern that forecasters may ironically delay decisions in anticipation of "just one more update," and in so doing miss the window within which a warning might have been ideally issued. Experimental research on this issue is greatly needed to determine the ideal update rate for decision making.
4. *Good heuristics, bad heuristics?* There is a general sense that when scholars evoke the term "heuristics," it is with a general tone of derision. Heuristics are discussed at length in the literature, as well as in this volume, but little attempt is made to develop a catalogue of heuristics forecasters generally use to make decisions. Thus, we are unsure which heuristics add value to objective forecasts, which compromise objective forecasts, and which have very little effect overall on the reliability and accuracy of predictions. We are also unsure about the social processes through which heuristics emerge within the forecasting office, and

when certain heuristics do emerge, why they become institutionalized and the processes through which they become so.

5. *Critically assess the role of the forecaster in an increasingly automated workplace.* Despite the weather's complexity, advances in mathematics, engineering, computer modeling, and radar technology make forecasting error increasingly less common. As a result, the forecaster may in the future play a diminished role in the forecasting process, and a greater role in interpreting weather information for institutional and public use. To account for the changing role of the forecaster, we should begin training and research programs on the best ways in which to communicate weather information to the public and the best practices for doing so.
6. *Understand and minimize stress.* Further understanding of the role of stress in the decision-making process for EMs and forecasters. A great deal has been done in other areas, but not severe weather warnings. As noted in Chap. 4, stress may create "tunnel vision" in decision making.
7. *Encourage the development of social bonds between EM and NWS.* Informal social relationships may play a key role in the development of understanding and trust between institutions. An emergency manager with network connections with National Weather Service personnel may dispense with many of the formalities of interaction in order to obtain direct, actionable information necessary to manage the weather emergency. Social bonds also establish direct network ties with relevant personnel. An emergency manager who needs a specific piece of information—such as radar data—knows specifically whom to contact if relationships are pre-established.

Bibliography

Baumgart, L. A., Bass, E. J., Philips, B., & Kloesel, K. (2008). Emergency Management Decision Making during Severe Weather, *Weather and Forecasting*, *23*(6), 1268–1279.

Betsch, T., Brinkmann, B. J., Fiedler, K., & Breining, K. (1999). When prior knowledge overrules new evidence: Adaptive use of decision strategies and the role of behavioral routines. *Swiss Journal of Psychology/Schweizerische Zeitschrift für Psychologie/Revue Suisse de Psychologie, 58*(3), 151–160.

Bond, N.A. and Mass, C.F. (2009). Development of Skill by Students Enrolled in a Weather Forecasting Laboratory. *Weather and Forecasting*, 24(4), 1141–1148.

Brotzge, J. and Donner, W. (2013). The Tornado Warning Process: A Review of Current Research, Challenges, and Opportunities. *Bulletin of the American Meteorological Society*, 94(11), 1715–1733.

Burger, J, Gochfeld, M., Jeitner, C., Pittfield, T. and Donio, M. (2013). Trusted Information Sources Used During and After Superstorm Sandy: TV and Radio Were Used More Often than Social Media. *Journal of Toxicology and Environmental Health, Part A*, 76(20), 1138–50.

Carter, H., Drury, J., James Rubin, G., Williams, R. and Amlôt, R. (2014). Emergency responders' experiences of and expectations regarding decontamination. International Journal of Emergency Services, 3(2), 179–192.

Crossman, A. n.d. "The Emergent Norm Theory and How It Affects Collective Behavior." *ThoughtCo*. Retrieved December 26, 2020 (https://www.thoughtco.com/emergent-norm-theory-3026305).

Demeritt, D., Cloke, H., Pappenberger, F., Thielen, J., Bartholmes, J. and Ramos, M-H. (2007). Ensemble predictions and perceptions of risk, uncertainty, and error in flood forecasting. Environmental Hazards, 7(2), 115–127.

Demeritt, D, Nobert, S., Cloke, H. and Pappenberger, F. (2010). Challenges in Communicating and Using Ensembles in Operational Flood Forecasting. *Meteorological Applications*, 17(2), 209–22. doi: https://doi.org/10.1002/met.194.

Demeritt, D., Stephens, E.M., Créton-Cazanave, L., Lutoff, C., and Nobert, S. 2014. Communicating and Using Ensemble Flood Forecasts in Flood Incident Management: Lessons from Social Science." In: Duan Q., Pappenberger F., Thielen J., Wood A., Cloke H., Schaake J. (eds) Handbook of Hydrometeorological Ensemble Forecasting. Springer, Berlin, Heidelberg. https://doi.org/10.1007/978-3-642-40457-3_44-1

Demuth, J.L., Morss, R.E., Morrow, B.H., and Lazo, J.K. (2012). Creation and Communication of Hurricane Risk Information. *Bulletin of the American Meteorological Society*, 93(8), 1133–45. doi: https://doi.org/10.1175/BAMS-D-11-00150.1.

Donner, W., Granger, D., Rodriguez, H., Diaz, W., and Santos, J. (2005). "Rainfall Estimates or Tornado Detection?: An Assessment Based on the Needs of Emergency Managers." ERC-CASA End-User Integration Research Brief No. 4, Disaster Research Center, University of Delaware, Newark, DE.

Donald H. and Wohl R. (1956). Mass Communication and Para-Social Interaction. *Psychiatry*, 19:3, 215–229.

Dorner, D. and Pfeifer, E. (1993). Strategic Thinking and Stress. *Ergonomics*, 36(11), 1345–60. doi: https://doi.org/10.1080/00140139308968004.

Doswell III, C. (2001). "Severe Convective Storms--An Overview." *Meteorological Monographs*, 28(50), 1–26. doi: https://doi.org/10.1175/0065-9401-28.50.1.

Doswell III, C. (2004). "Weather Forecasting by Humans--Heuristics and Decision Making." *Weather and Forecasting*, *19*(6), 1115–1126. doi: https://doi.org/10.1175/WAF-821.1.

Doswell III Charles & Schultz David (2006). On the use of indices and parameters in forecasting. Electronic Journal of Severe Storms Meteorology. 1. 1–22.

Drury, J., Novelli, D. and Stott, C. (2013). "Psychological Disaster Myths in the Perception and Management of Mass Emergencies." *Journal of Applied Social Psychology*, 43(11), 2259–70. doi: https://doi.org/10.1111/jasp.12176.

Edland, A and Svenson, O. (1993) Judgment and Decision Making Under Time Pressure. In: Svenson O., Maule A.J. (eds) Time Pressure and Stress in Human Judgment and Decision Making. Springer, Boston, MA.

Ernst, S., D. LaDue, and A. Gerard, 2018: Understanding emergency manager forecast use in severe weather events. Journal of Operational Meteorology, 6 (9), 95-105, doi: https://doi.org/10.15191/nwajom.2018.0609

Fischer, H.W. 1998. *Response to Disaster: Fact Versus Fiction & Its Perpetuation: The Sociology of Disaster*. University Press of America.

Funk, T. W. (1991). Forecasting Techniques Utilized by the Forecast Branch of the National Meteorological Center During a Major Convective Rainfall Event, *Weather and Forecasting*, *6*(4), 548–564.

Gladwin, T.N., Kennelly, J.J., and Krause, T-S. (1995). Shifting Paradigms for Sustainable Development: Implications for Management Theory and Research. *Academy of Management Review*, 20(4), 874–907. doi: https://doi.org/10.5465/amr.1995.9512280024.

Goldsmith, Barry S., Mike Buchanan, Scott Cordero, and John Metz (2012). "The use of Innovative Communications to enhance weather Information for a Rare Winter Storm In South." 92nd American Meteorological Society Annual Meeting, New Orleans, LA.

Heinselman P., LaDue D., Kingfield D.M., and Hoffman, R. (2015). Tornado Warning Decisions Using Phased-Array Radar Data. *Weather and Forecasting*, 30(1), 57–78.

Hoffman, R. R., LaDue, D. S., Mogil, H. M., Roebber, P. J., & Trafton, J. G. (2017). *Minding the weather: How expert forecasters think.* Cambridge, MA: MIT Press.

Horton, D., & Wohl, R.R., (1956). Mass Communication and Para-Social Interaction: Observations on Intimacy at a Distance. Psychiatry, 19 (3): 215–229. https://doi.org/10.1080/00332747.1956.11023049.

Hoss, F. and Fischbeck, P. (2015). Increasing the Value of Uncertain Weather and River Forecasts for Emergency Managers. *Bulletin of the American Meteorological Society*, 97(1), 85–97. doi: https://doi.org/10.1175/BAMS-D-13-00275.1.

Howard G R. (1986). The effects of time pressure on judgment in multiple cue probability learning, Organizational Behavior and Human Decision Processes. *Organizational Behavior and Human Decision Processes*, 37(1), 83–92.

Johnson, V. (2020). Designated Market Areas (DMAs) and Severe Weather Coverage: Oklahoma-Based Broadcast Meteorologists and Their Decisions for Providing Live, On-Air Coverage of Tornado Warned Storms. MA Thesis, University of Oklahoma, Norman, OK.

Joslyn, S.L., and Grounds, M.A. (2015). The Use of Uncertainty Forecasts in Complex Decision Tasks and Various Weather Conditions. *Journal of Experimental Psychology: Applied,* 21(4), 407–17. doi: https://doi.org/10.1037/xap0000064.

Kahneman, D and Tversky, A. (1973). On the Psychology of Prediction. *Psychological Review*, 80(4), 237–51. doi: https://doi.org/10.1037/h0034747.

Kuster, C.M., Heinselman, P.L., Snyder, J.C., Wilson, K.A., Speheger, D.A., and Hocker, J.E. (2017). An Evaluation of Radar-Based Tornado Track Estimation Products by Oklahoma Public Safety Officials. *Weather and Forecasting*, 32(5), 1711–26. doi: https://doi.org/10.1175/WAF-D-17-0031.1.

League, C.E., Díaz, W., Philips, B., Bass, E.J., Kloesel, K., Gruntfest, E., and Gessner, A. (2010). Emergency Manager Decision-Making and Tornado Warning Communication. *Meteorological Applications*, 17(2), 163–72. doi: https://doi.org/10.1002/met.201.

Maule A.J. and Svenson O. (1993) Theoretical and Empirical Approaches to Behavioral Decision Making and Their Relation to Time Constraints. In: Svenson O., Maule A.J. (eds) Time Pressure and Stress in Human Judgment and Decision Making. Springer, Boston, MA.

McCarthy, S., Tunstall, S., Parker, D., Faulkner, H. and Howe, J. (2007). Risk Communication in Emergency Response to a Simulated Extreme Flood. *Environmental Hazards*, 7(3), 179–92. doi: https://doi.org/10.1016/j.envhaz.2007.06.003.

Miller, R.C. (1972). *Notes on Analysis and Severe-Storm Forecasting Procedures of the Air Force Global Weather Central.* Volume 200 of Technical report (United States. Air Weather Service).

Novak, D. R., Bright, D. R., & Brennan, M. J. (2008). Operational Forecaster Uncertainty Needs and Future Roles, *Weather and Forecasting*, *23*(6), 1069–1084.

Petak, W. (1985). Emergency Management: A Challenge for Public Administration. *Public Administration Review, 45*, 3–7. doi:https://doi.org/10.2307/3134992.

Quarantelli, E.L. (1991). Lessons from research: findings on mass communication system behavior in the pre, trans, and postimpact period of disasters. PRELIMINARY PAPER #160, Disaster Research Center, University of Delaware, Newark, DE.

Rodríguez, H., Diaz, W., Donner, W., Santos, J., and Marks, D. (2005). *Allocation of Radar Resources and Policy Implications: The End-User Community in Oklahoma.* ERC-CASA End-User Integration Research Brief No. 3, Disaster Research Center, University of Delaware, Newark, DE.

Ramos, M. H., Van Andel, S.J., and Pappenberger, F. (2013). Do Probabilistic Forecasts Lead to Better Decisions? *Hydrology and Earth System Sciences*, 17, 2219–32. doi: https://doi.org/10.5194/hess-17-2219-2013.

Roe, A. (1951). A psychological study of physical scientists. *Genetic Psychology Monographs, 43 (2)*, 121–235.

Roebber P.J., Schultz D.M., Colle B.A., Stensrud D.J. (2004). Toward Improved Prediction: High-Resolution and Ensemble Modeling Systems in Operations. *Weather and Forecasting*, 19(5), 936–949.

Roebber, P.J., and Bosart, L.F. (1996). The Contributions of Education and Experience to Forecast Skill. *Weather and Forecasting*, 11(1), 21–40. doi: https://doi.org/10.1175/1520-0434(1996)011<0021:TCOEAE>2.0.CO;2.

Rogell, R. H. (1972). Weather Terminology and the General Public. *Weatherwise*, 25(3), 126–32. doi: https://doi.org/10.1080/00431672.1972.9931588.

Rothstein, H. G. (1986). The effects of time pressure on judgment in multiple cue probability learning. Organizational Behavior and Human Decision Processes, 37(1), 83–92. https://doi.org/10.1016/0749-5978(86)90045-2

Savelli, S., and Joslyn, S. (2013). The Advantages of Predictive Interval Forecasts for Non-Expert Users and the Impact of Visualizations." *Applied Cognitive Psychology*, 27(4), 527–41. doi: https://doi.org/10.1002/acp.2932.

Sherman-Morris, K. (2005) Tornadoes, television and trust—A closer look at the influence of the local weathercaster during severe weather. Global Environmental Change Part B: Environmental Hazards, 6:4, 201-210, DOI: https://doi.org/10.1016/j.hazards.2006.10.002

Sherman-Morris, Kathleen & Poe, Philip & Nunley, Christopher & Morris, John. (2020). Perceived Risk, Protective Actions and the Parasocial Relationship With the Local Weathercaster: A Case Study of Hurricane Irma. Southeastern Geographer. 60. 23-47. https://doi.org/10.1353/sgo.2020.0003.

Simmons, K. M. and Sutter, D. (2009). False Alarms, Tornado Warnings, and Tornado Casualties. *Weather, Climate, and Society*, 1(1), 38–53. doi: https://doi.org/10.1175/2009WCAS1005.1.

Smith, M.R. (2002). "Five Myths of Commercial Meteorology." *Bulletin of the American Meteorological Society*, 83(7), 993–96. doi: https://doi.org/10.1175/1520-0477(2002)083<0993:FMOCM>2.3.CO;2.

Spiegler, D.B. (1996). "A History of Private Sector Meteorology." Pp. 417–41 in *Historical Essays on Meteorology 1919–1995: The Diamond Anniversary History Volume of the American Meteorological Society*, edited by J. R. Fleming. Boston, MA: American Meteorological Society.

Stokes, C., and Senkbeil, J.C. (2017). Facebook and Twitter, Communication and Shelter, and the 2011 Tuscaloosa Tornado. *Disasters*, 41(1), 194–208. doi: https://doi.org/10.1111/disa.12192.

Stuart, N. A., Schultz, D. M., & Klein, G. (2007). Maintaining the Role of Humans in the Forecast Process: Analyzing the Psyche of Expert Forecasters, *Bulletin of the American Meteorological Society*, *88*(12), 1893–1898.

Suess, E., Cervato, C., Gallus, W., and Hobbs, J. (2013). Weather Forecasting as a Learning Tool in a Large Service Course: Does Practice Make Perfect? *Weather and Forecasting*, 28, 762–771. doi: https://doi.org/10.1175/WAF-D-12-00105.1.

Svenson, O., & Maule, A. J. (Eds.). (1993). *Time pressure and stress in human judgment and decision making*. Plenum Press.

Tierney, K., Bevc, C., Kuligowski, E. (2006). Metaphors Matter: Disaster Myths, Media Frames, and Their Consequences in Hurricane Katrina. *The ANNALS of the American Academy of Political and Social Science*, 604(1), 57–81.

Uccellini, L. W., Kocin, P. J., Schneider, R. S., Stokols, P. M., and Dorr, R. A. (1995). Forecasting the 12–14 March 1993 Superstorm, *Bulletin of the American Meteorological Society*, *76*(2), 183–200.

Weaver, J., Harkabus, L., Braun, J., Miller, S., Cox, R., Griffith, J., & Mazur, R. (2014). An Overview of a Demographic Study of United States Emergency Managers. *Bulletin of the American Meteorological Society, 95*(2), 199–203.

Wenger, D.E. (1985). "Mass Media And Disasters." Preliminary Paper #98, Disaster Research Center, University of Delaware, Newark, De.

Wilson, K.A., Heinselman, P.L., Kuster, C.M., Kingfield, D.M., and Kang, Z. (2017). Forecaster Performance and Workload: Does Radar Update Time Matter? *Weather and Forecasting*, 32(1), 253–74. doi: https://doi.org/10.1175/WAF-D-16-0157.1.

Wright, P. (1974). The Harassed Decision Maker: Time Pressures, Distractions, and the Use of Evidence. *Journal of Applied Psychology*, 59(5), 555–61. doi: https://doi.org/10.1037/h0037186.

Chapter 5
Towards a Model to Coordinate Interdisciplinary Weather Research

Interdisciplinary collaboration between the natural and social sciences is, in relative terms, a fairly new enterprise within general efforts of scholarly inquiry. Indeed, formal collaboration between scientists from different natural science disciplines only became commonplace mid-century, and, at that, slowly. While prior to the early twentieth century some scattered interdisciplinary research took place, multidisciplinary collaboration was far from common, with the emergence of fields such as "biophysics, biochemistry, and bioengineering" arriving as late as the 1950s (Klein 1990: 32). In an historical account of interdisciplinarity, Klein remarks that "disciplinary cooperation was not unknown in the natural sciences, but a 'fissiparous tendency' dominated the infrastructure of science until approximately mid-century" (23). As the United States entered the 1960s, however, radical shifts in the organization of scientific research had taken firm hold and moved toward reshaping the theory and practice of scholarship, resulting in a willingness among academics to increasingly traverse disciplinary boundaries in the pursuit of basic and applied scientific innovation. As society grew in both size and complexity, researchers of the time increasingly acknowledged the need for interdisciplinary approaches (Hagstrom 1964). Although much of this collaboration continued to involve exclusively multidisciplinary efforts among natural scientists, scientific problems were

---

Interdisciplinarity holds a variety of different meanings depending on context. For instance, collaboration between two natural scientists (for instance, a computer scientist and geneticist) or interdisciplinary degrees (e.g., women's studies). Oftentimes, particularly in the context of interdisciplinary degrees, "interdisciplinarity" means "looking at a topic from multiple disciplinary perspectives." The interdisciplinarity we discuss in the present volume, while naturally incorporating multiple perspectives and stakeholders, goes further in the effort to reconcile methodologies between the natural and social sciences. For the purposes of the book, and while interdisciplinarity holds a variety of meanings, "interdisciplinarity" here is treated as collaboration between natural scientists (both heavily theoretical as well as the applied sciences, including engineers) and social scientists in efforts to develop synthesized approaches to solving real-world problems. Importantly, also note that we use "natural science" as a catch-all reference for any STEM-related discipline, including, importantly, engineering.

W. Donner et al., *Severe Weather Warnings: An Interdisciplinary Approach*,
https://doi.org/10.1007/978-3-031-05031-2_5

increasingly regarded as far too complex to be solved by one single discipline, thereby forging the path for collaborations between the natural and social sciences—as well as the humanities—in the decades that would follow.

Historically, while there is little possibility in identifying where interdisciplinary research between the natural and social sciences began, we know that the National Science Foundation's Engineering Research Centers (ERCs) presented the first institutionalized settings in which collaboration variously between STEM disciplines and the social sciences would first formally originate, incubate, and evolve. Implemented in 1985 as an NSF initiative, the concept and philosophy of the ERC, developed in collaboration with the National Academy of Engineers, was twofold in intent: first, to improve the quality of engineering research and education and, relatedly, to assist the United States in remaining competitive in an increasingly globalized economy (NSF n.d.). In the spirt of these goals, first-generation ERCs centered principally on manufacturing engineering with the objective of developing market competitiveness across the U.S. industrial sector. Much of these earlier efforts saw collaboration confined among engineering subdisciplines—for instance, civil engineering and computer engineering—rather than between engineering and the social sciences. As time progressed, however, a growing acknowledgement of the systemic nature of engineering—that engineered systems are part of a human social system—gave way to a second phase of ERC development from roughly 1994–2006, within which time a broader definition of interdisciplinarity was adopted in projects such as *ERC for Quantum Energy and Sustainable Solar Technologies, several Engineering Earthquake Research Centers (EERCs), and ERC-CASA* (Preston and Lewis 2020). It is worth highlighting that within this time, notes Preston and Lewis, "the phrase 'engineering systems' was changed to 'engineered systems'" (n.p.) in what amounted to implicit acknowledgement of the need for a more nuanced conceptualization on the nature and character of innovation within the applied sciences. This evaluation in values and thinking on the matter is perhaps best reflected in language from a 1998 Program Announcement: "An *engineered system* (authors' emphasis) is derived from integrating a number of components, processes, and devices to perform a function. The system may be living or inanimate in origin. It must be complex and challenging enough to justify a ten-year program of research. Analysis, modeling, or development of the individual components of a system, without their integration into a complex engineered system, is not an appropriate focus for an ERC" (NSF 1998: 3). While the shift in language did not explicitly call for increased collaboration between the natural and social sciences, nevertheless, this marked a significant break from the past in which engineering systems were considered exclusively in terms of technical features and components. This all would suggest that technical innovation was increasingly viewed as a necessary but ultimately insufficient element of the engineering project—the innovation, as it came to be gradually thought, must also through design fulfill some *useful purpose*.

Yet, today, collaboration between the natural and social sciences as a model for the practice of interdisciplinary research remains an elusive problem for which no present solution exists. Indeed, there is at present no generally accepted model for multidisciplinary work, nor any common roles through which the common needs of an interdisciplinary project can be met and addressed. As a result, there is limited guidance on how to set up and maintain a large, long-term interdisciplinary effort. Recall the previous chapters provided a state-of-the-art assessment of public response to hazard warnings, emergency management and forecaster decision making, and the factors that shape scientific thought and protocol with respect to severe weather threats. In the present chapter, we develop a model through which to develop and organize a large-scale interdisciplinary weather project. The dynamics of warning systems—fundamentally governed by complex interactions between environment, technology, and society—ranks as perhaps one of the clearest examples of the need for collaboration across multiple fields of study.

Several questions naturally attend the development of a general model of interdisciplinary research. How should such a project be organized? What are the key roles and stakeholders involved in such an undertaking? How do we distinguish meaningful interdisciplinary questions? How might problems and issues of collaboration, trust, and accountability be resolved? And perhaps most importantly: what makes an interdisciplinary project fundamentally interdisciplinary? To proceed, we highlight and outline two important features of the organized interdisciplinary endeavor. First, we ask: what *is* interdisciplinary research between the natural and social sciences? What *defines* the interdisciplinary approach? What are some common problems associated with interdisciplinary research, and, accordingly, does the literature provide any guidance in solving these problems? Further, how do we apply a general interdisciplinary perspective to the context of warning systems, and what are the needs of the various subcommunities in the process and the philosophy of interdisciplinary warning design? In examining these questions, we give less attention to the conceptual and theoretical literature on interdisciplinary research, favoring, rather, a deeper focus on the empirical literature in order to provide practical guidelines for collaboration. Here we ask and address: "What establishes and maintains collaboration between team members, particularly among those from highly unrelated disciplines (e.g., sociology and meteorology, political science and chemistry, etc.)?" "What builds a practical context of collaboration, encourages teamwork across the disciplines, and ultimately cultivates a sustainable, ongoing scientific partnership?" "What leads to measurable outcomes of success in an interdisciplinary context?" It is the intention of the next two chapters to outline what might be termed a "best practices" approach to developing an interdisciplinarity project, with concrete recommendations based upon (a) tested models of interdisciplinary collaboration and (b) empirical studies of interdisciplinary teams and teamwork. We hope to articulate the model in such a general way to make it an effective starting point for any interdisciplinary project, weather-related or otherwise, so that it might serve as a useful guide for scholars and practitioners.

To be sure, there are many different kinds of interdisciplinarity. A sociologist may work with a psychologist to study, for instance, the influence of group dynamics on some psychological outcome. Alternatively, a computer scientist may work with a biologist to develop computer models of animal behavior. In this chapter, however, it must from the beginning be emphasized that we are interested in interdisciplinary work between social scientists, on the one hand, and applied natural scientists, on the other hand, the latter including meteorology and engineering. Much of the work done in this regard is based on the philosophy of an *end-to-end* system design in which *end-users*, those for whom systems are immediately useful or beneficial, are included as active stakeholders and research subjects in the scientific development of novel technologies. End-users come from a variety of backgrounds and are unique to the subject matter of an interdisciplinary project. In the case of severe weather warnings, of which is principal concern for this volume, end-users may be regarded as, but not limited to, emergency managers, forecasters, private industry, mass media, and the general public.

Distinctions Between the Natural and Social Sciences as a Starting Point

The struggle to develop a workable definition of interdisciplinarity likely originates from the philosophical, theoretical, and methodological contrasts—some real, others merely attributed—between the natural and social sciences. Each regards the nature and dynamics of human psychology, society, and culture in fundamentally different ways. Although natural science and engineering are often cast as "objective" and "value-free," both, it is argued, share unspoken—and as a consequence oftentimes untested—assumptions about human thought and behavior with respect to systems. This has led to two specific problems for the formulation of interdisciplinary approaches. First, STEM-related fields seem to take for granted the presence of an end-user who will always and under all circumstances employ a technical system as intended by the designers. As was seen in previous chapters, this is an unrealistic assumption. Humans maintain a tendency to "break" the intended rules and expectations of purely engineered systems, which is to say that such systems will more likely than not clash with or defy the way end-users will go on to use the system in practice. In designing a new warning system, for instance, it is perhaps natural that engineers may assume an instinctual and reflexive response to risk on the part of people and build such assumptions into system design. This is perhaps not terribly surprising since U.S. culture—of which U.S. engineers and scientists are part and into which they are socialized—places heavy value on individualism and puts forth the belief that the individual is, at the end of the day, the ultimate arbiter of his or her own decisions. Deterministic sociological forces that subvert free will, be they of social, cultural, or psychological origins, are therefore generally not conceived as fundamentally relevant to the warning process by those who would design such systems. Nor does the highly specialized character of modern science encourage reflection beyond one's immediate field—and this goes for social scientists, as well. However, as was seen in Chaps. 3 and 4, these assumptions and practices must be called into question. We know that a system without a user context, no matter how well engineered from a scientific and technical standpoint, will be of limited value to users. The point of this discussion is that although presenting itself as impartial, natural science tends to naturally and inevitably "build in" to system design specific assumptions about users—namely, that end-users are rational and calculating, applying some form of pure logic and reason in the use of the system that somehow inevitably aligns with the intention of the designer.

Illustrating these general points, Catney and Lerner (2009) discuss at length their experiences on an interdisciplinary sustainability project referred to as SUBR:IM (Sustainable Brownfield Regeneration: Integrated Management), which involved revitalizing "brownfields"—former industrial or commercial zones in which there may be substantial contamination— in the United Kingdom. The project quickly developed an interdisciplinary perspective as the research involved the potential for redevelopment to create the condition for changes in demographics, inflation of housing prices, and the potential for sprawl, as the authors note, which lie at the nexus of the "complex array of decision-making processes and interactions" (295). Discovered were what Catney and Lerner labeled "conceptual immensurability" between natural and social scientists working on the project. It was noted that while natural scientists thought in terms of the metrics of the pollution control as the ultimate evaluation of sustainability, the social scientists regarded as central to sustainability "social equity, effective governance or interpretation and actualization of the idea by particular groups, such as planners and property developers" (302). Thus, the nature of the problem and the context of the problem differed between both groups as the matters most central to the project were regarded in different terms.

Second, methodological differences also account for difficulties in any collaborative effort intending to bridge disciplinary boundaries. The standards and practices of sampling, data collection, and analysis within the natural and social sciences are often presented as incompatible. While not entirely inaccurate, the contrast, although present, is nevertheless very much overstated and to a large degree imputed. Nuijten (2011) regards the distinction as a false dichotomy, arguing that the basic epistemologies—that is, positivism and realism—undergirding the methodologies of the natural and social sciences establish at the very least the form a common philosophical approach to inquiry. At the same time, challenges emerge because the social sciences also rely a great deal on qualitative research, an approach less in accord with the dominant, quantitative approach of natural science. This is because qualitative research avoids characterizing phenomena in purely discreet, numerical terms as does the strictly quantitative approach of natural science. Perhaps most importantly, qualitative research attempts to communicate an understanding of the context in which human behavior takes place, a goal for which quantitative research is unsuited by design and by character. For example, a quantitative survey asks interviewees to report the number of false alarms experienced in the past month. For this approach to be successful, however, the assumption must hold that all respondents hold the same general definition of a false alarm. If reported false alarms are based on unique definitions of false alarms between interviewees, then the reported number is in effect meaningless. A qualitative researcher, on the other hand, may ask, "Can you please describe to me what a false alarm and whether you ever take them into account when deciding to respond? If so, in what ways?" Here there is no assumed definition of a "false alarm." Rather, the research is interested in how interviewees *themselves* define a false alarm, how they arrive at this definition, and how it impacts their decision making (if at all). All of this simply cannot be captured by a quantitative survey since such an approach assumes that there is a stable definition around which a consensus in thinking among interviewees exists. It is for this reason, therefore, that the qualitative approach of the social sciences and the dominant quantitative approach of the natural sciences are largely responsible for this perceived distinction between the two. In particular, integrating the narrative elements of qualitative research into the predictive models of natural science has been an ongoing challenge for interdisciplinary efforts. Common ground, however, has not been forthcoming.

A Note on Definition and Terminology

There is presently no generally-agreed-upon conceptual or applied framework for interdisciplinary research (Pennington 2011). In one of the earliest attempts at conceptualization, the Committee on Facilitating Interdisciplinary Research, Committee on Science, Engineering, and Public Policy (2004) defined interdisciplinary research "as a mode of research by teams or individuals that integrates information, data, techniques, tools, perspectives, concepts, and/or theories from two or more disciplines or bodies of specialized knowledge to advanced fundamental understanding or to solve problems whose solutions are beyond the scope of a single discipline or area of research practice" (2). Other definitions attempt to establish conditions that call for an interdisciplinary approach. Tait and Lyall (2007: 2) summarize five criteria by which the suitability of a question for interdisciplinary research might be evaluated:

- the nature of the subject is interdisciplinary (e.g. transport, environment)
- researchers are transferring information from the laboratory to the real world
- the research is user driven (not necessarily commercial)
- the research is particularly relevant to policy making in complex areas
- single discipline research has encountered a bottle-neck and more than one discipline needed to make a breakthrough.

Still others have attempted to create conceptual taxonomies categorizing the different forms interdisciplinary research might take. Stock and Burton (2011), for instance, provide one such system of classification: *disciplinary, interdisciplinary, multidisciplinary, and transdisciplinary*. These categories are purported to differ in fundamental ways. Some allegedly possess the ability to blend observation and theories into new disciplines, others are focused on solving real-world problems and coordinating multiple researchers and stakeholders, whereas others specialize in cultivating "pluralist methodologies" (1101). While useful conceptually, oftentimes these typological definitions are so vague as to be of little service in a practical context, lacking an ability to distinguish in a meaningful, pragmatic way between interdisciplinary and monodisciplinary research. The problem is that we have not uncovered, to date, the very basic definition of interdisciplinarity, much less one that is useful to applied research. Both interdisciplinary research and monodisciplinary research, for example, solve real world problems, incorporate multiple stakeholder groups, and coordinate multiple researchers. None of these are unique to either approach. Moreover, the elements that do indeed distinguish the four typologies of interdisciplinary research remain difficult to capture conceptually or operationalize practically. What does it mean, for instance, to construct a "pluralist methodology?" There is little consensus regarding how one does so in practice, or, for that matter, what a "pluralist methodology" might look like in a laboratory. In order to develop a model for interdisciplinary research, a common, useful, and easily-definable element must be identified. In response, we simply apply the term *interdisciplinary research* to describe that state and process of scientific teamwork in which more than one discipline through collaboration conducts research on some unresolved question that holds consequences for society. Beyond this, no further distinctions are made as they are not regarded as useful in the present discussion. We must establish and define interdisciplinary research in practice moving ahead, which is to ask: *what makes a question itself interdisciplinary, or, what is interdisciplinarity in its practical form?*

The Social Scientist in Interdisciplinary Research

An extended example might serve to better clarify precisely how the interdisciplinary approach synthesizes natural and social scientific efforts. Here we attempt to illustrate how, as mentioned in the previous section, interdisciplinary research might look "in practice." Before proceeding, it should emphasized that a large-scale multidisciplinary project often involves not a single, grand collaboration across the disciplines, but rather a collection of interrelated subprojects in which representatives from various disciplines take part cooperatively. On a single severe-weather project, for instance, sociologists might collaborative with engineers to study tornado lead time, psychologists might collaborate with forecasters to develop psychologically-resonant warnings, and economists might work with civil engineers to balance cost with safety in the design of public shelters.

What, then, do social scientists do in the context of interdisciplinary research? To answer this question, it must first be understood that natural scientists and engineers in designing human-relevant systems[1] make inescapable assumptions—almost always implicitly—about the people who are to use or in some way benefit from the existence of the proposed system. Engineers, for instance, design technical elements of warning systems (or any system, for that matter) under specific and oftentimes very narrow presumptions about how people think and behave. Indeed, sometimes the assumptions of engineers are correct; other times, the ways in which

[1] Perhaps a redundant phrase as all engineered systems are, alas, "human-relevant systems," but we include to avoid any risk of ambiguity.

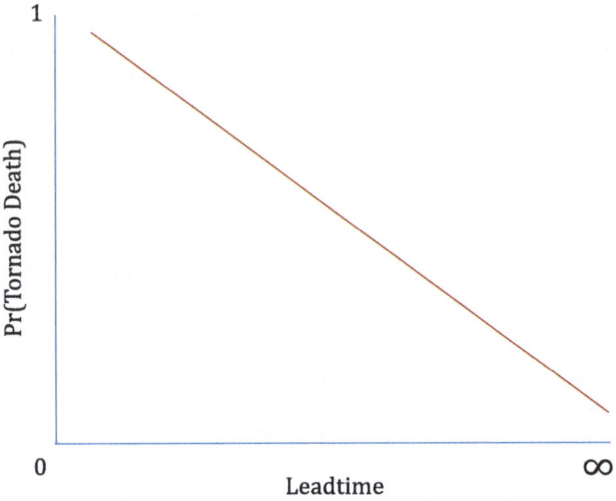

Fig. 5.1 Assumed indirect, linear relationship between warning leadtime and tornado mortality

technologies are anticipated to be used and how they are used *in situ* diverge in unexpected and sometimes dramatic ways. Interdisciplinary research intends to test the assumptions of engineered systems so that people may enjoy full benefit, thereby bridging the gap between design and context. Accordingly, *the social scientist's foremost role in an interdisciplinary project is to translate into* testable, hypothetical statements *the underlying assumptions that a novel engineering system implicitly makes about human thought and behavior*. This is perhaps the most simple and above all practically useful way in which to conceptualize interdisciplinary research and yet general enough to be portable from project to project.

This is perhaps excessively abstract. To illustrate this argument more concretely, here we use an example with which the authors are familiar: collaborative research in the field of remote sensing and radar engineering. In the following hypothetical illustration, a group of radar engineers attempt to develop a novel radar design system able to detect and predict severe weather (e.g., tornadoes, floods, severe thunderstorms) more accurately and reliably than was possible in the past due to preexisting limitations in weather sensing technologies. One of the principal goals of the new system might be to increase severe weather "lead time," which is operationally defined as the length of time between the time at which a severe weather event is identified and the time at which the warned event strikes the community (recall Chap. 3 for a more in-depth discussion of lead time). Tornado lead times have historically ranged from between 5 to 15 min. Project engineers hope the proposed radar system will, under ideal circumstances, make possible the extension of lead times into hours, if not days. Such a design, while seemingly rational, implies a far too uncomplicated assumption about warning response psychology: namely, that terror and terror alone motivates people to remain sheltered until authorities lift warnings. In graphical terms, this results in an equally simple model specifying an inverse, linear relationship between lead time and safety in which longer and longer lead times result in fewer deaths (See Fig. 5.1) as people are assumed to remain sheltered for the duration of any lead time length due to the overwhelming and

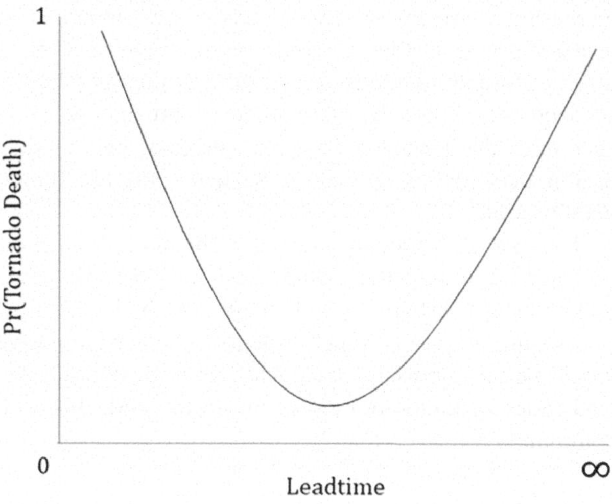

Fig. 5.2 Hypothesized "U-shaped" relationship between warning lead time and tornado mortality

irresistible compulsion of fear. Accordingly, under this premise, it logically follows that the system should be designed in such a way to extend lead times indefinitely—hours, perhaps days—if the technology could be designed sufficiently to do so. But this premise is grounded in several assumptions that may or not be true in the context of how people actually behave under periods of warning. It assumes, for instance, that people will *tolerate* sheltering indefinitely (if indeed they choose to shelter, which is unlikely in a number of contexts). That they will not get bored or become anxious. That they will not leave the shelter prematurely for any number of reasons after a given period of time and subsequently find themselves exposed to the event. These counter assumptions call into question the wisdom of developing systems capable of extending lead times indefinitely. And here is the central, unavoidable point of criticism brought forth by interdisciplinary research: *each of these assumptions—which are really testable hypotheses—must hold true if the original engineering approach is to be the correct one, if the approach is to bring forth what is ultimately sought: a useful function.* It is the role of the social scientist to determine which of these assumptions are correct and provide guidance to the engineers accordingly.

Regarding the proposed system, the social scientists involved in the project confer with the engineers and then go about interpreting and converting into testable hypotheses the assumptions made about human behavior within the current design approach. In our hypothetical example, data on community response, lead times, and epidemiology are collected by the social scientists, and results suggest patterns somewhat different from those assumed in the engineering model. Instead of a straightforward inverse linear relationship, what emerges from our hypothetical research is rather a "U-shaped" curve between lead time and tornado mortality (See Fig. 5.2). As lead time increases from "0 min," there is a marked decline in fatalities. As lead times is observed to increase further, however, the line begins to curve upward, suggesting at some point increasing lead time (contrary to expectations) holds the opposite effect: longer lead times beyond a certain point lead to increasing numbers of deaths. Under the imagined scenario, further qualitative research suggests that, at some point, shelter-seekers under very prolonged lead times eventually come to regard the warning as false, emerge from shelter, and are subsequently killed by the tornado immediately or attempt to flee via car and encounter the tornado. Thus, the original behavioral assumptions and resulting design decision to develop a system with indefinite lead time is found to be inappropriate.

The role of the social scientist in this hypothetical example would be to provide guidance regarding the "ideal" lead time. In this case, the ideal lead time is not indefinite, but perhaps around 15 min, in which case the community has enough time to respond, but not so much time to endanger themselves by driving off. Indeed, it would be an impressive technical feat to design a radar system that could provide lead times of hours and days, but, in the end, this would not serve to protect the community.

A General Organizational Model for Interdiciplinary Research

For the purposes of beginning a broader discussion on the topic of interdisciplinarity, here we present a general model of interdisciplinary research. Discussed here are the common roles within an interdisciplinary effort and recommended organizational structures best suited for interdisciplinary research.

Common Roles

Within the severe weather interdisciplinary effort, there are several generally acknowledged roles (See Table 5.1 and Fig. 5.3). As a scientific enterprise, naturally included among these roles are *STEM scholars* tasked to address specific, technical questions such as, for example, "how do we innovate radar technology and practice to ensure more reliable predictions of severe weather", or "what building materials best withstand the extreme violence of an earthquake or water pressure?" Providing social, psychological, and cultural context to these raw technical questions are *social scientists* who might ask, for instance, "what are the socioeconomic barriers to the affordability of new earthquake-proof materials" or "how will a radar system's improved accuracy integrate with present emergency management doctrine and institutions?" Finally, there are a heterogenous group of *end-users* or *stakeholders* for whom the technical innovation is beneficial or useful in some regard.

End-users may fall into three broad groups and merit a bit of additional discussion. While all end-user and stakeholder groups play different parts and occupy different roles within an interdisciplinary project, some end-users share a more dynamic role than others. Some end-user groups may be the object of study and,

Table 5.1 Development of a hypothetical severe weather warning project

Field	Problem	Solution
Meteorology	• Prediction of severe weather • Detection of severe weather	• Improved algorithm accuracy and validity • Clearer or more specific visualization formats
Engineering	• Radar optimization	• More sophisticated radar technologies
Computer Science	• Algorithms for detecting severe weather	• Code and algorithm refinement
Social Science	• Warning reception • Warning response	• Culturally- and socially-relevant communication • Understand response needs of diverse communities (e.g., norms, language, SES, etc.) • Overcome barriers to believing and understanding warnings • Role of emotion and cognition in processing warnings

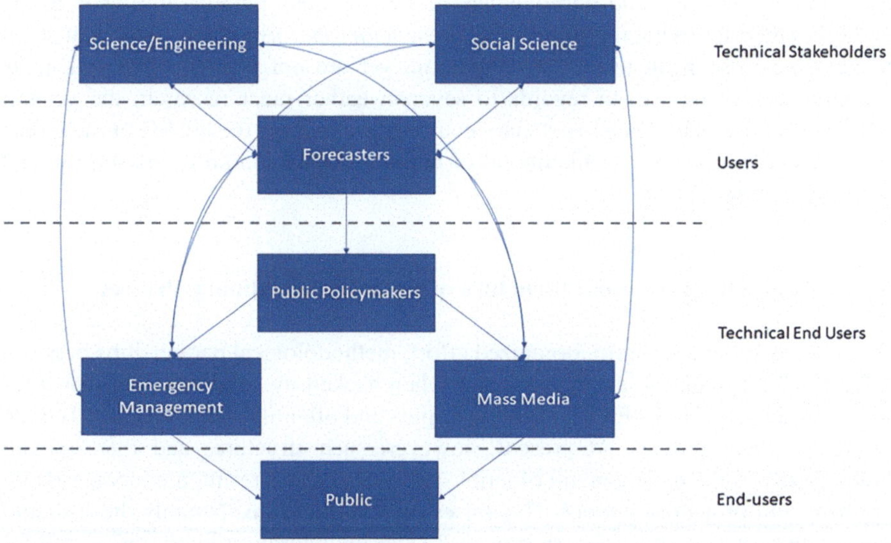

Fig. 5.3 End-to-end design of warning systems. The figure above illustrates a "model" network for the organization of a severe-weather end-to-end research project. Note that a majority of the arrows are two-way, suggesting the need for routine interaction and feedback between the various scientists and stakeholders involved in the project

at other times within the same project, collaborate with scientific team members on research or policy. Meteorological forecasters serve as an obvious example in this regard. On the one hand, an interdisciplinary project may seek to understand forecaster decision making in the context of a proposed radar innovation. On the other hand, the same forecasters may work with the scientific team members to optimize radar data or craft public surveys of warning response in collaboration with social scientists. This leads to yet a further related distinction—that between core end users, mid-range end users, and beneficiary end users. *Core end-users* may be defined as those end-users who more-or-less will directly apply technical innovations developed through the project. Continuing with the previous example, this would be a role likely played by meteorological forecasters on a radar engineering project. Forecasters may make use of more accurate radar data to distinguish more reliably, for instance, between tornadic and non-tornadic storms. What separates this group from the *mid-range end-users* is that core end-users more often than not hold scientific training likely comparable to the scientific researchers and engineers involved in the project, and, for that reason, will commonly take part in and advise the development of the scientific approach, as well. *Midrange end-users* oftentimes share many of the practical tasks of core end-users, but lack the formal scientific training and experience to actively develop and test theoretical questions alongside the core scientific team. Emergency managers, for instance, fall into the category of mid-range end users. By way of example, emergency managers and forecasters may both use and interpret weather data, but forecasters are likely better equipped to assist in the collection and improvement of weather data. Finally, at the end of the end user chain lies what might be called *beneficiary end-users*: end users that stand to gain the most from technical innovation, yet are not actively involved in its scientific design or use. In regards to severe weather radar research, the general public may be considered a beneficiary end-user group since the quality of radar data has a direct bearing on the likelihood of severe weather mortality, morbidity, and property damage.

Towards an Organizational Structure of the Interdisciplinary Project

As with any other long-term, organized effort, methodological partnerships between STEM and the social sciences work best when backed by administrative structures and roles designed to fulfill the highly-complex and oftentimes dynamic needs of an interdisciplinary project. At present, however, such structures and roles remain defined only in the most general of terms and, perhaps as a result, are inconsistently applied from project to project. This raises several questions. Notably, how should we organize our interdisciplinary projects? Does the interdisciplinary vision call for a novel set of roles and functions the need for which is mostly absent in monodisciplinary projects? If so, what are these roles and functions, and how do these roles change the overall organizational and relationship structure of interdisciplinary efforts, and what might be a starting point for developing an organizational model for collaboration between STEM and the social sciences? Long-term interdisciplinary projects, especially those comprised of multiple, diverse disciplines with a history of limited collaboration, require a clear pathway for design and execution above and beyond what might be demanded from monodisciplinary research.

This is because there are several notable uncertainties, questions, and problems unique to interdisciplinary research. For example, how do we translate methodologies, theories, and ideas across the disciplines in a way that is sensible and relevant?

How are internal and external stakeholders (i.e., core, mid-range, and beneficiary end-users) meaningfully integrated into the design and execution of the project? How is progress audited and assessed? How do we ensure that collaboration remains a consistent feature of the project from the start and not merely implemented in an *ad hoc* fashion? To summarize: what are the specific needs of an interdisciplinary project, and from that, what are the key roles needed to address those needs? As a complex undertaking, the organization of interdisciplinary research is unlikely to be simple and straightforward. Indeed, an interdisciplinary effort is likely to require great care and complexity of design in order to meet the challenges posed by collaboration between fields whose methodologies and theories oftentimes stand at great variance.

Rather than reinvent the wheel as it were, here we draw upon and adapt preexisting organizational models in an effort to provide guidance on the design of a large-scale interdisciplinary project (with the strong caveat that the following should serve as just as that—a guide, adaptable to the needs of individual projects, rather than a set of rigid and inflexible set of rules applied without the consideration of context). First, as is with science in general, it must be understood that long-term interdisciplinary research increasingly operates in bureaucratic terms, and, as is the case with bureaucracies in general, requires therefore well-defined roles and structures in order to operate effectively and efficiently. An interdisciplinary project may involve multiple collaborating institutions, both public and private, composed of different, oftentimes highly-specialized groups within which and between which hierarchies form. There are managers, administrators, scientists, administrative assistants, postdocs, students, and a host of external stakeholders, all of whom must be managed and engaged in the project to varying degrees. Budgets must be built, maintained, and accounted for. Goals and mission statements drafted. A general order and momentum must be maintained.

From this, there can be little doubt that interdisciplinary research in practice has needs similar to a modern complex organization and therefore requires similar roles to address these needs. Thus, our search for useful administrative roles for an interdisciplinary project naturally begins in the field of administrative science, which suggests that the universal needs across bureaucracies include **mentorship**, **facilitation**, **innovation**, and **coordination** of project activities (Quinn 1988). On an interdisciplinary project specifically, König et al. (2013), in applying Quinn's model to collaborative research, suggests that the mentor and facilitator role work towards establishing and maintaining an *interdisciplinary culture* while innovator and coordinator are critical for *end-user engagement* (both scholars refer to this as *open systems*, but we here call this simply "end user engagement" for the sake of conceptual simplicity). These functions respectively work towards the necessary goals of ensuring meaningful, ongoing interdisciplinary collaboration between project members in a complex, bureaucratic environment, while, at the same time, building external relationships with core, mid-range, and beneficiary end-users to develop support networks and translate project findings in a lay vernacular. This ensures that the project remains focused on collaboration.

Establishing and Maintaining an Interdisciplinary Culture

According to König et al., as far as the "interdisciplinary culture" function is concerned, the *mentor* and *facilitator* roles have an unambiguous task: cultivating an interdisciplinary identity among interdisciplinary project personnel as well as formalizing the principles and expectations of multidisciplinary collaboration in order to "create a common vision to work towards as well a common language to communicate needs among the different work groups" (König et al. 2013: 266). In terms of mentorship, this involves "concern for different needs of the project team…building commitment…[and] maintaining morale, whereas the facilitator function involves the duties of providing means for discussion and participation, while maintaining openness" (266). A notable task is the translation of disciplinary concepts across disciplines and interdisciplinary concepts to the project team as a whole. For instance, the facilitator may communicate and explain sociological concepts and methodologies to the team of meteorologists and vice-versa.

Even with a stated commitment to the interdisciplinary vision, members may naturally gravitate towards ideas, methodologies, and theories with which they are familiar and comfortable—again because of the highly specialized nature of modern science. Thus, ongoing interdisciplinary projects face the ongoing challenge of maintaining inertia in order to avoid decomposing into a collection of more-or-less discrete disciplinary projects without a common focus on overall goals. This calls for a need to routinely build and maintain a sense of common identity, arrange structures that foster collaborative teamwork, and mediate disagreements among team members when the need arises. It is arguably the goal area most critical to the interdisciplinary effort. Building upon Quinn's (1988) work, König et al. (2013) identifies two main roles necessary to maintain a culture of collaboration—the *mentor* and the *facilitator*. Both roles should be occupied by persons familiar with interdisciplinary research, preferably scholars who have been involved with interdisciplinary research in the past. The *mentor* supports big-picture project cohesion by cultivating a collaborative framework in order to prevent a "siloing" effect that may naturally arise and divide teams composed of researchers from highly-specialized backgrounds. The project mentor would be tasked to continually nurture an interdisciplinary focus, assist members with interdisciplinary ideas, and find ways to build an interdisciplinary inertia back into the project if such a siloing effect is present. Mentors work to counteract the almost natural decentralizing tendency common to the organized interdisciplinary endeavor.

The *facilitator*, under König et al., plays the far more "in-the-trenches" role of translating for STEM and social science ideas and concepts that may be foreign to either group. Whereas the mentor possesses a more general responsibility of keeping the project interdisciplinary in nature, the facilitator holds a much more technical and specific duty in making the scientific work itself interdisciplinary and intelligible to researches and stakeholders. As noted earlier, oftentimes STEM scholars and social scientists speak past one another in theoretical and methodological terms—indeed, one can rightly conceive of each speaking in fundamentally different languages.

Thus, here the facilitator has the profoundly important role of translating natural science ideas for social scientists and social science ideas for natural scientists, in so doing attempting to bridge the methodological and philosophical gulfs discussed earlier. Of all tasks, however, among the most critical is translating the assumptions of STEM and engineering design approaches into testable hypothetical statements about human behavior in collaboration with social scientists. The need for this arises because, once again, STEM-based and engineering design decisions include implicit hypothetical statements about human behavior that must be tested and evaluated to determine whether the assumptions—and, by implication, the design decision itself—hold. As such, the facilitator role is perhaps the most critical and important among all the roles in the project because the facilitator bears the responsibility of making the project ultimately useful to stakeholders and end-users, and in doing so ensuring that the system will be truly useful to those for whom it is intended—not the designers, but communities and society in general. The person best suited for such a role, accordingly, would ideally be a scholar with broad training in the social and natural sciences, and, ideally, experience working on past interdisciplinary projects. This has been a challenge for most interdisciplinary projects as there is a lack of dedicated interdisciplinary training within the University system to produce scholars theoretically and methodologically dynamic enough to work in both worlds and bridge the divide on projects whose members arrive from diverse academic backgrounds.

End-User Engagement

End-user engagement specialists—what both Quinn (1988) and König et al. (2013) refer perhaps too abstractly as *open systems*—reflects the need to adapt goals to stakeholder and end-user needs. Here the *innovator* and *broker* roles take a central part in the engagement, inclusion, and management of stakeholders. König et al. (268) describes these roles within the context of interdisciplinary research: "[the innovator provides] insight into user needs and their translation towards adaptation within the research context to develop novel solutions," whereas the broker serves as a facilitator of the "iterative exchange between the funding body, the potential users of the developed tools and methods, and the researchers." All things external to the project are dealt with by the end-user specialists. Importantly, König et al. remarks that members of the user engagement team hold the important job of communicating and marketing scientific ideas, results, and applications to non-scientists and non-scientific institutions. Both roles typically engage what were earlier termed mid-range and beneficiary end-users through the development of techniques to disseminate, translate, and market project findings to a broader lay audience, some of whom may be more directly involved in the project than others. Within the context of an interdisciplinary warning project, these groups may include, but are not limited to, the general public, industry, public safety institutions, local government officials, fire, private weather corporations, and others. For the innovator, a strong background in social science—sociology, psychology, political science, anthropology, etc.—is necessary in order to develop methodologies through which to measure and assess end-user needs. For the *broker* role, strong public relations skills, as well as, ideally, a familiarity with the stakeholders to whom the project is relevant, is necessary to encourage *systematic* buy-in and long-term engagement and partnership. The broker, for instance, may work with industry and industry leaders to form partnerships. Ideally, funding permitting, a formally-trained public relations representative would be employed in both the broker and innovator roles. To summarize more generally, while the interdisciplinary culture function ensures that the project is internally integrated, the "open systems" functions ensures that the project is externally aligned with those who will use and benefit from the system in some way. The next chapter details some of the problems, duties, and responsibilities these various roles encounter on an interdisciplinary project.

Conclusion

The preceding chapter has been an attempt to call attention to the need for the development of a model for interdisciplinary research. Provided was an overview of the history of social science research, a distinction between natural and social sciences, a model for identifying interdisciplinary roles, and, finally, the establishment of roles needed to support the large-scale, organized interdisciplinary effort. Now, moving to Chap. 6, we move on to the common problems encountered in an interdisciplinary project and accordingly offer a rough outline or blueprint for the development of collaborative research between the natural and social sciences.

Bibliography

Committee on Facilitating Interdisciplinary Research, Committee on Science, Engineering, and Public Policy (2004). Facilitating interdisciplinary research. National Academies. Washington: National Academy Press, p. 2.

Hagstrom, W. O. (1964). Traditional and modern forms of scientific teamwork. Administrative Quarterly, 9(3), 241–263.

Klein, J. T. (1990). *Interdisciplinarity: History, theory, and practice.* Detroit, MI: Wayne State University Press.

König, B., Diehl, K., Tscherning, K. and Helming, K. (2013). "A Framework for Structuring Interdisciplinary Research Management." *Research Policy,* 42(1), 261–72. doi: https://doi.org/10.1016/j.respol.2012.05.006.

NSF. (1998). Program Announcement, Engineering Research Centers, Partnerships with Industry and Academe for Next-Generation Advances in Knowledge, Technology, and Education (NSF: 98-146). Directorate for Engineering, National Science Foundation, p. 4.

NSF. n.d. Creating New Knowledge, Innovators, and Tehcnologies for Over 30 Years. https://www.nsf.gov/eng/multimedia/NSF_ERC_30th_Anniversary.pdf, Accessed 8/1/20.

Nuijten, E. (2011). Combining research styles of the natural and social sciences in agricultural research. NJAS: Wageningen Journal of Life Sciences, 57(3–4), 197–205

Pennington, D. (2011). Collaborative, cross-disciplinary learning and co-emergent innovation in eScience teams. Earth Science Informatics. 4. 55–68. https://doi.org/10.1007/s12145-011-0077-4.

Catney, N. and Lerner, D.N. (2009) Managing Multidisciplinarity: Lessons from SUBR:IM, Interdisciplinary Science Reviews, 34:4, 290–308, DOI: https://doi.org/10.1179/030801809X12529269201129

Preston, L. and Lewis, C. (2020). Agents of Change: NSF's Engineering Research Centers--A History. Potomac, MD: Lynn Preston Associates LLC.

Tait, J. and Lyall, C. (2007). Short Guide to Developing Interdisciplinary Research Proposals ISSTI Briefing Note (Number 1) March 2007

Stock, P. and Burton, R.J.F. (2011). Defining Terms for Integrated (Multi-Inter-Trans-Disciplinary) Sustainability Research. *Sustainability*, 3(8),1090–1113.

Quinn, R.E. (1988). *Beyond Rational Management: Mastering the Paradoxes and Competing Demands of High Performance*. San-Francisco, CA: Jossey-Bass

Chapter 6
Building and Troubleshooting an Interdisciplinary Project

We now shift from a proposed system of structures and roles of an interdisciplinary project to a discussion of common problems associated with an interdisciplinary project. The roles described earlier—*mentor, facilitator, innovator, and broker*—play a significant role in this troubleshooting process. Described here are the normal—though far from exhaustive—set of problems that most commonly beset the long-term interdisciplinary project. As noted earlier, many of the issues and problems typically seen throughout the course of interdisciplinary research result from the presence or absence of an *interdisciplinary culture*, the establishment of which involves principally efforts of the project's mentors and facilitators. Accordingly, issues related to trust building, collaboration, diagnosis of engineering assumptions, as well as other themes, all of which related to the degree to which a cross disciplinary effort is made, will be explored in the following section. It will be seen that interdisciplinary researchers confront a variety of unique challenges in the planning and execution of an interdisciplinary project, the success of which requires collaboration between STEM and the social sciences. To be sure, some of the problems common to collaborative research are not necessarily unique to collaborative research *per se*—the acquisition of funding; the selection of appropriate sampling, data collection, and analytic methodologies; and developing an overall theoretical lens through which to channel efforts remain part and parcel of any investigation regardless of disciplinary focus. And, in this sense, interdisciplinary research is not fundamentally different from disciplinary research. So here we focus mainly on the challenges confronted uniquely by interdisciplinary researchers. As will be seen, collaboration between STEM and the social sciences—particularly collaboration across the disciplines—erects distinct obstacles above and beyond those encountered among projects conducted through a purely disciplinary strategy.

W. Donner et al., *Severe Weather Warnings: An Interdisciplinary Approach*,
https://doi.org/10.1007/978-3-031-05031-2_6

Establishing a Global Question and Transforming Applied Scientific Questions into Social Science Hypotheses

For a time, we have known that some questions are simply too complex to be addressed by a single discipline (Klein 1990). This view is far more accepted today than it has been in the past. Yet there is still uncertainty about what interdisciplinary research truly is. In other words, we lack a model for developing truly interdisciplinary questions. This is because interdisciplinary research is often described in deeply abstract and intangible terms. Interdisciplinarity, in common parlance, is defined as an "innovative ecosystem" within which team members develop "synthetic methodologies" and discussions of new theories leading "paradigm shifts" in innovation. But, in purely practical and technical terms, what does collaborative research ultimately boil down to, and how do we do it in a way that might justifiably be called interdisciplinary? The most useful way to orient and pursue interdisciplinary research may be to turn an eye towards what Haapasaari et al. (2012) have called the "global question," or the problem towards which all researchers, irrespective of disciplinary focus, channel efforts in collaborative terms. What distinguishes the "global problem" from ordinary science and engineering questions (e.g., how to improve radar data reliability) is through an orientation towards making system innovation useful to people and not merely innovative from an exclusively technological standpoint. However novel and advanced, improvements to existing systems must be useful to people, as well. Does it save lives? Does it make people healthier? Does it protect their property and homes? Does it provide some measure of security? These are global questions around which interdisciplinary research should ultimately center and towards which all sub questions and subprojects should channel. Once again, going back to an earlier discussion, we may be able to develop radar systems capable of extending lead time to hours or days (a tremendous technical innovation to say the least), but the wisdom of doing so is questionable given what is known about public response and human psychology. Thus, with the concept of a global question in mind, the general focus of engineering reorients from simply "how do we increase lead time?" to the global question: "how do we save lives?" In reframing the question globally, wherein lives saved—rather than simple lead time length—is a metric for success, we are in turn compelled to rethink the basic utility of lead time (or any other technical improvement, for that matter) in reference to the global question: "will extending tornado lead time to hours or days save lives?" This is an empirical question, and one that can be answered through social science research, which in turn informs whether the effort to develop a system intended to provide longer and longer lead times is indeed sound and therefore necessary—for, again, *longer lead times may not save lives, they may simply be longer*.

Again, the need for a global question emerges from what is the most basic tenet of interdisciplinary research, which we once again revisit for emphasis: *the fundamental need for an interdisciplinary approach extends from the reality that applied questions of natural science and engineering automatically formulate unspoken hypotheses about human psychology and behavior even if such assumptions remain formally undeclared.* It is through the establishment of a global questions that members of a project take the first steps towards determining what these assumptions are, which is why identifying and agreeing upon global questions from the outset is so fundamental to the success of an interdisciplinary project. At the end of the day, whether one labels collaborative research "multidisciplinary" or "transdisciplinary," or goes on to discuss purely abstract matters related to "methodological syntheses" or "pluralist methodologies," this stands as the defining feature *sine qua non* of the interdisciplinary approach: to look for the social science assumptions within scientific statements and transform them into testable hypotheses, the results of which can then be fed back into system design decisions. This is what makes research truly interdisciplinary.

How might this look further in practice? Once again, it is important to bear in mind that the global question better synthesizes the overall goals of the STEM researchers with that of the social science researchers and therefore avoids being irrelevant to end-users. For instance, in the context of an interdisciplinary warning study, we might establish as the global question something to the effect of "How do we minimize risks associated with severe weather?" or "how do we design and develop a warning system with the express goal of saving lives?" A poor question from the perspective interdisciplinary research, on the other hand, may be along the lines of "how do we design a warning system with longer lead times" or "how do we design a radar with more precise imagery?" To reiterate, and this must strongly emphasized, the latter questions are naturally *important*, but poorly conceived from an interdisciplinary standpoint because they reduce the solution of severe weather risks to that of strict innovation. The founding of a global question, on the other hand, has an important function within an interdisciplinary project: it helps remind participants to looks carefully for unmet assumptions about human behavior within scientific statements and test those assumptions accordingly.

How might we move from the global question to a series of questions that address both the scientific and social scientific dimension of problems? The most general approach might include:

1. Articulate the immediate engineering strategy
    (a) What are the goals of the engineering strategy?
    (b) What is the reason for adopting this strategy in particular?
    (c) How will it address the global problem?
2. From this, state or summarize, as simply as possible, the engineering approach to the question or problem
3. Therein, articulate the assumptions made about human behavior
4. Transform assumptions into testable hypotheses which either confirm the engineering strategy or alter it

The advantage of such a general, pragmatic definition of interdisciplinarity research is that it avoids the problem of excessive conceptualization discussed earlier. *Accordingly, interdisciplinarity may be thought of and defined in its most general character as the movement from a global questions, deduced as engineering solutions, and further refined through social science evaluation*. Here we provide an example:

Global Question or Global Problem Statement: Reduce mortality associated with tornadoes

Solution 1

Stated engineering approach: Extend lead time as long as possible.

STEP 1: Identify implicit assumptions

Unstated social science assumptions: There is a linear relationship between lead time and response, i.e., longer lead times make people safer.

STEP 2: Turn assumptions into testable hypotheses

Testable Social science hypotheses: "If lead time is longer, there will be fewer fatalities"

STEP 3: Revision of stated engineering approach

Solution 2

Stated engineering approach: "Provide emergency manager with more radar data during periods of severe weather"

STEP 1: Identify implicit assumptions

Unstated social science assumptions: There is a direct, linear relationship between the amount of radar data and the quality of forecaster and emergency management decision making.

STEP 2: Turn assumptions into testable hypotheses

Testable social science hypotheses:

- "The use of radar data among emergency managers yields fewer false alarms or missed events at the local level"
- "The addition of radar data to the collection of tools used in decision making will not affect the speed with which decisions are made"
- "Emergency managers understand the nature and applications of radar data"

STEP 3: Revision of stated engineering approach

Solution 3

Stated engineering approach: Minimize the false alarm rate through improved radar technology

STEP 1: Identify implicit assumptions

Unstated social science assumptions: Fewer false alarms lead to greater future warning credibility and fewer fatalities

STEP 2: Turn assumptions into testable hypotheses

Testable social science hypotheses:

- "If one experiences a higher rate of false alarms, one is more likely to regard future false alarms with increasing skepticism"
- "Areas with higher false alarm rates experience fewer deaths"
- It may also be necessary in this case to examine how the public defines false alarm as the perceived or subjective false alarm rate may differ from the true or objective false alarm rate.

STEP 3: Revision of stated engineering approach

In closing, the establishment of a global question further works towards reducing conflict and disorder within an interdisciplinary project. Without a global question in mind, engineering approaches risk evolving into what Mitroff and Emshoff (1979) call "ill-structured (e.g., ill-defined, messy) organizational problems," (1) which occur when

> (a) The problem is well-defined in the sense that it can be clearly stated but those charged with dealing with it cannot agree upon an appropriate solution or strategy; (b) they cannot agree on a methodology for developing such a strategy; or (c) they cannot even agree on a

clear formulation (definition) of the problem (objectives, controllable variables, and uncontrollable variables) (1).

While Mitroff and Emshoff used the notion of ill-structured problems to make sense of the complexity that naturally accompanies the practice of business management, the reality that some problems exist without a clear, stipulatory methodological approach to resolution is perhaps the most central barrier to a successful interdisciplinary project. Applied scientific and engineering questions, as a means to resolve problems, are therefore by definition ill-structured, for they often do not take into account the implicit psychological and sociological questions discussed earlier. Again, all assumptions—either stated or implied in the engineering approach—should be identified and discussed, then the assumption, importantly, should be then rearticulated as a testable, hypothetical statement that may either be accepted or rejected.

Complexity and Loss of Focus on the "Global Question"

Interdisciplinary research is characteristically an ambitious affair, rife with complexities, ambiguities, and no small amount of uncertainty. Oftentimes these issues evolve over the life course of the collaborative effort however well-designed and well-intentioned the project. Consequently, one problem that emerges from this complexity is a the considerable risk that the data and information generated from the project may grow to such a volume and to such as scope as to become unmanageable. McFadden et al. (2009) cautions that interdisciplinary projects have a distinct tendency to yield far too much information relative to available resources, which may, as a result, jeopardize efforts at effective project governance. Given these concerns, it is therefore natural to ask: how big is too big and how complex is too complex for the interdisciplinary endeavor? As with any project, interdisciplinary projects may either be too ambitious from the outset, or, alternatively, grow at an unsustainable rate. Interdisciplinary projects, as a result, tend to be vulnerable to problems associated with organizational complexity when compared to projects involving only one discipline.

This is for a variety of reasons. Interdisciplinary efforts typically comprise a larger and more varied collection of researchers, groups, institutions, and organizations from across a broader spectrum of intellectual and scholarly diversity. Stakeholder affiliates also tend to be drawn from a broader, deeper pool, as well. For the average interdisciplinary project, such complexity may manifest in poorer outcomes overall. In an analysis of 62 NSF interdisciplinary research projects from 1998 to 1999, Cummings and Kiesler (2005) report on the factors that most frequently lead to the success or failure of projects conducted collaboratively through support by the Computer and Information Science and Engineering Directorate's (CISE) "Knowledge and Distributed Intelligence" (KDI) program. Following a post-workshop survey, data on the experiences of project team members were collected and assessed in order to understand the challenges associated with research grounded in diverse scientific teamwork. Notably, projects took place in a wide variety of contexts, some conducted within a single university, while other projects in the sample were conducted across multiple, diverse institutional settings. Speaking to the point of organizational complexity, Cummings and Kiesler found that larger, more distributed projects fared worse than projects with simpler, more centralized structures (again, this may be related to the fact that more frequent face-to-face interaction is associated with better interdisciplinary outcomes, as discussed earlier). Practices associated with project success, and which seemed to mitigate problems of complexity, included the frequency of face-to-face meetings, lab meetings, recruitment of external speakers, and in-person seminars, once again pointing to the efficacy of direct interaction in interdisciplinary project work. Workshops, conferences, or travel helped to somewhat further offset the problems associated with more complex interdisciplinary projects. Thus, interdisciplinary projects should include a manageable number of partners in as central a location as possible. If the nature of the project unavoidably calls for greater organizational complexity, however, the problems associated with more broadly distributed teams might be mediated somewhat through policies that encourage regular, face-to-face interaction among members as the literature would suggest.

Trust and Legitimacy Building Across Disciplines

As a result of specialization within the academy, little commerce has historically taken place between natural science and social science for most of the twentieth century, which has in part contributed to an ongoing belief that the two are, in a basic sense, incompatible in terms of methodology and world view. In an metanalysis of 81 papers dealing with interdisciplinary themes, paradigm differences between the natural and social sciences were identified by interviewees as a particularly troublesome aspect of the collaborative effort in roughly one-fifth ($n = 20$) of cases (Fischer et al. 2011). Thus, within any interdisciplinary project, an effort must be made to bring together, develop, and support a community of diverse scholars between whom significant barriers to trust are more likely than not to exist. In order to do so, project members must share a fundamental confidence and understanding of the interdisciplinary vision, in both a general sense as well as one related specifically to the project at hand. Gathering a group of diverse scholars together and asking them to collaborate simply will not do. Each must understand one another's contributions to the global question.

How then is understanding and trust built and sustained within an interdisciplinary work environment? Harris and Lyon (2013) observed specifically the process of trust building among research collaborators, exploring the processes through which confidence in interdisciplinary research is built and sustained, through which seeking to better understand (a) what motivates collaboration, (b) what deters collaboration, and (c) what factors cultivate trust among interdisciplinary group members. From a population of 53 collaborative studies, a sample of 10 were selected with varying combinations of disciplinary culture, partnership structures, and interaction styles. The final sample included a diverse range of interdisciplinary partnerships, including interdisciplinary teams from biosecurity to animal welfare to arable farming. Eight representative members of each of the ten projects were selected for interview. Results from the interviews suggest that trust building is a process driven by several factors, summarized by the authors as "having information on others, prior experience of working together, norms of cooperation, and sanctions exerted on those who might transgress norms of behavior. Having incentives is a necessary but not adequate basis for having trust. These relationships are built up through existing relationships, building trust through progression of projects and the use of intermediaries or guarantors" (109). As far as trust building goes, these findings seemingly call attention to the importance of working with the same team of scholars or researchers from project to project when possible, or at the very least including members who had at some point worked together in the past. This would appear to echo the conclusions of earlier research, which suggested that ongoing, sustained interaction plays a significant role in trust building, with informal networks in particular playing critical roles in successful projects, especially under circumstances where resources are absent or significantly restrained (McFadden et al. 2009). Naturally, this is not to preclude new members (far from it), especially ones with necessary expertise, or who might bring much needed inclusion and diversity to the project, but rather, according to research, encourage the recruitment of members who have worked together well in the past, or, in Harris and Lyon's (2013: 116) terminology share preexisting "norms of cooperation." In this way, interdisciplinary projects do not "start from scratch," and new members will have an important source of support and guidance from experienced interdisciplinary researchers. Ideally, when possible, there should be some fluency of interaction between team members from the start, which can be further cultivated by the project mentor and facilitator.

Face-to face interaction may not only build trust where none presently exists, but also serves to ease existing distrust between the disciplines or persons. Due to prevailing stereotypes and misunderstandings about the social sciences, natural scientists and engineers may lack confidence in methodologies underlying social scientific investigations, which may ultimately damage collaborative efforts. Research, however, shows that sustained, face-to-face interaction between natural and social scientists may serve to ultimately dilute such misconceptions over time. In a recent study, Kirby et al. (2019) find that earth scientists tend to believe, on average, that social science generally lacks the competency and rigor characteristic of natural science. However, returning to the point made earlier about the benefits of interaction between the disciplines, it was also shown that such negative attitudes among natural scientists tended to soften through ongoing collaboration with social scientists on research projects. Indeed, this is perhaps not surprising as the psychological literature suggests strong involvement and identification with those against whom negative attitudes are initially held has shown some success in minimizing stereotypes (Fitzgerald et al. 2019). Here this supports a point made earlier in the section: that confidence and trust within interdisciplinary partnerships does not simply come out of the blue, but is built and carefully tended through exposure and collaboration in the long-term.

Importantly, it is also noted in the literature that incentives alone may only go so far between members of collaborative research teams. In other words, carrots and sticks only work to an extent. Thus, in addition to providing incentives to collaborators, leadership must also work to craft an environment in which scholars simply *want* to collaborate or simply find value in the act of collaboration itself. Strong interdisciplinary partnerships do not develop overnight. Rather, the will to collaborate emerges and develops over time through sustained, successful collaboration, as well as through an acknowledgement of the mutual importance of disciplinary contributions, which, together, work to build trust and motivation for teamwork across the disciplines. Perhaps a concrete step towards achieving this goal is through a series of interdisciplinary retreats, the purpose of which serves to establish partnership frameworks and potential collaborative partnerships within the project. Retreats would also work to introduce all members to the concept and practice of interdisciplinary research, thereby preparing them for future work. Retreats are indeed an important mechanism here as research notes the importance of face-to-face social interaction in building trust, especially so in the context of interdisciplinary projects (Rhoten et al. 2009; Hampton and Parker 2011).

Engagement of Stakeholders

Successful and sustained engagement with stakeholders remains a feature critical to the success of an interdisciplinary project. Deverka et al. (2012: 185) provide perhaps the most straightforward and useful definition of research stakeholders: "Individuals, organizations or communities that have a direct interest int he process and outcomes of a project, research or policy endeavor." In an applied, interdisciplinary project, stakeholder engagement is paramount to success as a means of "setting priorities, funding programs, assessing applications, conducting research, and communicating findings," as well as generally assisting in the evaluation process (Boaz et al. 2018: 2). Stakeholders provide important information about the contexts into which technology will be introduced, establishing connections between the scientific team and beneficiary stakeholders, and also play an important role in "marketing" the technology to other stakeholder groups for whom the new technology is deemed relevant. While stakeholder engagement has been found to improve the quality of scientific findings and applicability of technologies mainly in the area of healthcare research (Boaz et al. 2018), stakeholder integration nevertheless has not historically been prioritized in interdisciplinary research.

For instance, in a 2005 study of 62 NSF interdisciplinary projects, Cummings and Kiesler (2005) found that the time and effort devoted to scientific discovery often far eclipsed time and effort devoted to the engagement of stakeholders, resulting in fewer than a third of the PIs reporting significant partnership with industry or the community at the time of project conclusion (note that stakeholder and end-user engagement receives far more emphasis, however, within NSF proposals at the present time). Since the outcome of interdisciplinary research is to improve some aspect of technology and society—again, returning to the notion of the "global question" discussed in the previous section—an interdisciplinary project without a sustainable and realistic plan for stakeholder and end-user partnerships may result in failure to capitalize on the most important dimension and strength of the interdisciplinary approach to scientific research. When a concerted effort towards stakeholder engagement is made, a key challenge then becomes identifying and managing the different interests of stakeholder participants. While stakeholders on an interdisciplinary project may take on a variety of roles, there are two that are perhaps most important: (a) *setting a research agenda* and (b) *knowledge translation*.

Due to previous experience, stakeholders tend to develop an understanding of problems and needs associated with their fields or circumstances, and therefore are critical to developing a comprehensive research agenda—what is relevant to and how to approach the study of the "global question" of the project. Regarding *research agenda setting*, Kohlhagen et al. (2013) reports on the critical role of stakeholder engagement within an interdisciplinary project on the environmental sustainability of inland swamps, noting the impact of both geophysical and anthropogenic factors on swamp quality, the latter of which including such factors as urbanization and swamp geographic proximity to urban areas. Here it is argued that since the influence of society and culture—notably, conditions related to urbanization—factored so heavily into swamp management, it follows that the presence of community stakeholders involved in the development of urban planning and policy was critical to the success of the project in both scientific and policy aims alike. As key stakeholders, urban planners might, for instance, provide advice and context for planned environmental engineering initiatives—that is, whether such initiatives accommodate past and future planning needs, and whether such approaches are appropriate for the setting. Planners, while themselves not formally scientists or researchers, as expert stakeholders nevertheless know and understand the problems of urban development and expansion and can therefore accordingly work with environmental scientists in developing relevant questions. Haney and Power (1996) remark that the process of stakeholder integration must occur from the very beginning of a project. In doing so, what Kohlhagen et al. (2013) refer to as a "baseline" for research is therefore established whereby researchers may determine relevant questions and avenues of inquiry relevant to the input and advice of stakeholder groups, thus, once again, working towards answering the "global question" towards which an interdisciplinary project is naturally oriented. Stakeholders may also play a significant role in advising researchers on how best to sample and collect data from subjects, thereby providing valuable methodological insights. Emergency managers, many of whom may be volunteer, may not be willing to travel great distances to take part in a focus group and therefore may need to be studied through an online survey, a fact that might be discovered through stakeholder groups on a warning project. Knowing this from the beginning (once again, establishing a "baseline" as recommended by Kholhagen) through consultation with local stakeholders, and being able to adjust the sampling strategy in response, improves the quality and efficiency of data collection immeasurably.

Another key activity of stakeholders is the *transfer and interpretation of knowledge* to other stakeholders in the project. Grimshaw et al. (2012) suggests that a significant challenge on a research project lies with what has come to be known as "knowledge translation," or the communication of relevant concepts and practices to stakeholders or end-users who may lack the formal expertise to understand scientific findings or ideas. This is an important concept naturally for an interdisciplinary project since the transfer of knowledge is critical to making innovation relevant and available to society. Accordingly, project researchers, while well versed in the design of research methodologies, may lack the ability or direct credibility to interpret findings in such a way as to resonate with certain stakeholder elements. Stakeholders, for instance, may need information communicated in a non-technical way. With this in mind, some stakeholder groups may take the role of knowledge translators for other stakeholder groups. For instance, in a warning project involving multiple stakeholder groups, weather forecasters may play the vital role of translating technologies and meteorological concepts for emergency managers. Why forecasters? Forecasters, once again, have the scientific expertise to understand project concepts, while, at the same time, know and understand the needs of emergency managers, and so serve as effective professional emissaries between core scientists and public institutions.

With these considerations in mind, the question then becomes: how does one engage stakeholders and project end-users in a way that is both systematic and comprehensive? Taking these considerations into account, the process may be distilled into three steps:

1. *Develop an exhaustive list of stakeholder individuals or groups*. To continue with a theme central to this volume, in an interdisciplinary tornado warning project the stakeholder structure is vast and complex, comprising numerous different groups, organizations, institutions, and communities, all of which and all of whom differ dramatically with respect to technical expertise, sustained interest in severe weather, disciplinary visibility, and anticipated level of engagement with the research project, project members, and the proposed technology (see Table 6.1). For instance, forecasters, on the one hand, as a stakeholder group, as far as training, experience, and expertise go, are in many ways indistinguishable from the scientists running the project (i.e., meteorologists, engineers, computer scientists, etc.). On the other end of the spectrum rests the general public, most of whom think little about the weather outside of times in which the likelihood of severe weather is heightened, likely, or imminent. Engaging these dramatically different end-user population requires distinct strategies.
2. *Distinguish between core and peripheral end-user groups, organizations, and institutions*. A plan for identifying and distinguishing between, as exhaustively as possible, each end-user group, their respective roles in the study (e.g., assisting core scientific teams, research subjects, etc.), and how they will be incorporated into the study must be developed. While end-user groups are dramatically

Table 6.1 End-user types and associated project roles

Group	Roles
Core end-users • Weather forecasters • National Weather Service (NWS) • Storm Prediction Center (SPC)	• Assist in experimental design • Provide high-level scientific guidance • Provide active, direct guidance to system design • Provide feedback on utility of system design • Translate knowledge and findings to lower-ranked end-users
Mid-range end users • Emergency managers • Floodplain managers • Government • Storm spotters • Volunteer organizations	• Assist in experimental design • Provide low-level pragmatic guidance • Liaison with Core-end users • Provide feedback on utility of system design
Beneficiary end-users • General public • Community institutions	• No or very limited role in system design • Typically end-stage end-user group for whom the system holds benefit • Typically studied

different, they nevertheless share general qualities in the context in interdisciplinary research that may serve as a starting point for categorization and engagement strategies. The first group are core stakeholders. Some of these groups actively engage in the research and development of project goals, contributing knowledge, expertise, and time in collaboration with the core team. Once again, forecasters fall into this groups in the context of interdisciplinary warning research. Others are "beneficiary" end-users. These end-users stand to benefit directly from the system and oftentimes have the most to gain or lose from the system itself. The public falls in this category since they are not involved in the design of severe weather technology, but are, in many ways, the most profoundly affected by said technology's state of development. In the middle once again fall what might be called "mid-range end users," or those groups for whom the system is of primary consequence to decision making and who may likely have some role or direct input into the design process, though to a far less extent in comparison to core end-users. Social scientists often study the decision making, judgements, and needs of mid-ranged end users to develop guidance to the core scientific team as new technologies are developed. This is so the technology meets the distinct need of the mid-range end user groups. For instance, on a severe weather interdisciplinary project, emergency managers oftentimes occupy the role of a mid-range end user in that they do not necessarily have the scientific expertise to develop the system, but nevertheless possess enough practical knowledge to actively guide system design and provide context to decision making in the design process. How emergency managers make decisions, their capacity for using multiple, quickly-evolving sources of data, level of training, and financial constraints are all relevant to how a radar system is designed and deployed and must be accordingly taken into account in the design process.

3. Finally, *categorize and plan engagement activities*. The next step is to categorize the end users according to the rubric under 2. Beneficiary end-users, with little input into the project, will likely not be engaged to the extent of the other groups mentioned. They are most likely to be the subjects of study rather than active partners in research. "Mid-range" end-users, emergency managers, for instance, may be the subjects of study as noted previously, but also offer insights into the research and planning process of the core scientific team. Finally, core-end users may be subjects of study as well, but, at the same time, are likely to contribute to project design and analysis. For instance, on a severe weather warning project, forecasters may assist meteorologists conducting experiments in the collection of wind or reflectivity data. All needs must be taken into account for the project to be successful.

Conflict Management

While conflicts may arise between members of a single discipline, here we discuss conflict management in the context of disagreements and misunderstandings that arise between members from different disciplinary backgrounds. Disagreements between STEM researchers and social scientists may extend from a variety of contexts, according to the literature. As Bracken and Oughton (2006) note, increasing specialization within STEM and social science fields has led over the years to the development of dissimilar vocabularies through which scholars have come to identify and address scientific questions, which has, in turn, to some extent frustrated attempts to find common ground between natural and social scientists in the execution of interdisciplinary projects. Without a common language through which to negotiate the development of theories, methodologies, and analytic strategies, collaboration of any form bears a tendency to stall. Barthel and Seidl (2017) indirectly support this notion by demonstrating that when successful collaboration between the natural and social scientists did indeed take place, such collaboration typically occurred between natural scientists and economists. This is likely due to the fact that econometric vocabularies and methods, which tend to be overwhelmingly quantitative, most closely resemble and emulate those of the natural sciences. Consequently, such collaborations may occur more naturally and with much less effort, and so the two fields may gravitate towards one another in the context of interdisciplinary research. This again speaks to the notion that fields sharing a common methodology and a common language may be prone to less conflict during periods of collaboration.

Indeed, research demonstrates communication and conflict as commonly present within interdisciplinary projects. In a survey of 323 academics from the social and natural sciences at universities, research centers, and grant organizations, Roy et al. (2013) observed that among the various problems respondents identified, communication issues on interdisciplinary research projects between natural and social scientists ranked among the highest. Perhaps related, also identified as potential sources of conflict were unresolved methodological disputes between the natural and social sciences. Institutional factors, such as a disregard for interdisciplinary work and publications, also featured heavily into the discussion of findings.

Another root cause of conflict, according to Lele and Norgaard (2005), relates to the distinct culture of natural science, on the one hand, and social science, on the other, that tend to run somewhat conceptually perpendicular in what has been referred to as "mismatched taxonomies" (969). Lele and Norgaard offer a further elaboration of this concept:

> Scientists working in a subdiscipline often tend to believe that their particular way of categorizing phenomena (taxonomic system) is the best way of characterizing reality, rather than being open to different ways of representing reality that might be more or less appropriate in different applied contexts. For instance, for years soil scientists have been debating as to which system of soil taxonomy is the best. The US Department of Agriculture (USDA) soil taxonomy system appears to have won the battle and become the most commonly accepted classification. But if the purpose of soil classification is to relate soil types to forest vegetation types or agricultural fertility, the USDA soil taxonomy is not very useful. Indeed, when soil scientists shift from working on fertility questions in agricultural soils to working on sustainability questions in forests, they may have to change not only their taxonomy but also their methods (typical soil depths sampled, parameters analyzed, etc.). But once the problem of mismatched taxonomies is recognized and addressed, communicating across disciplines in the natural sciences becomes quite easy. (969)

How should interdisciplinary conflict be addressed and resolved? Notably little attention has been devoted to conflict management between groups within an interdisciplinary project. Underacknowledged in this respect remains the potential for tension between the various disciplines involved in an interdisciplinary project for reasons which may include the identification of different goals, perceptions of human nature and character (as discussed earlier in Chap. 5), adherence to different values, and the reliance on different methodologies. What is more, status hierarchies between disciplines may also frustrate attempts at collaboration. Relative to the social sciences, the STEM fields have been regarded with greater prestige and more deserving of resources (Machlup 1988), a position which may favor the natural sciences with respect to the distribution of resources across the life of the interdisciplinary project. (This is not the say that the natural sciences should not receive more in resources—all else qual, natural science research is costlier—yet, at the same time, the social sciences may as a result be forced to operate on a minimal or shoestring budget that is incommensurate with methodological need. A budget must be allocated, again, not towards mere technical innovation, but rather towards the resolution of the global questions towards which all member work). Under such circumstances, the potential for conflict is not unlikely, especially in the absence of formal project structures and roles through which communication and understanding are encouraged.

Another fact that must be acknowledged is that the natural and social sciences are by and large tethered to sometimes overlapping, sometimes opposed, research protocols—from determining a meaningful unit of analysis, to sampling techniques, modes of observation, to analyses methods. Discussing his experience mediating a cross-disciplinary effort focused on desertification and land use in Southern Europe, Jeffrey (2003) reports several issues that arose during his tenure as the project's mediator (perhaps most akin to Koenig et al.'s "Mentor" role). The goal of the research was to examine decision making in the farming industry—more specifically, what factors influenced or determined decisions to irrigate pump water, windmill use, and fertilize were to be predicted in the model. Jeffrey (2003: 545) notes that in exploring the issue "two distinct, and contrasting, methodological systems are evident. . .the very nature of computer simulation imposes a structure on both the type of analysis and information that are to be used. . .in contrast, the knowledge possessed by social scientists (concerning farmers' perception of crop change) suggests at best a semi-structured process. . .[which is] opaque and the contributing considerations are often confused and variable across individuals." The fundamental problem in this sense, the solution of which is perhaps too abstract to solve in this volume, is how to reconcile and ultimately synthesize quantitative elements of computer modeling, which reduces human behavior to a single, numerical dimension, and a qualitative fieldwork project, which does the opposite by expanding the boundaries of the object of inquiry on the basis of human experience. The fundamental problem in this case comes down to the very different ways natural and social scientists represent the human condition and human behavior.

Rewards and Incentive Structures Across the Disciplines

A territorial barrier to the development of not only interdisciplinary projects, but also to the more general programmatic adoption of interdisciplinarity within the academy, remains a reluctance to acknowledge disciplinary publications outside of one's field or interdisciplinary publications in general. Scholars, especially junior scholars, may therefore be reluctant to risk publishing in journals beyond the scope of the discipline within the context of interdisciplinary research. During an annual review process, interdisciplinary journals may not be counted equally (if at all) along with traditional, disciplinary venues, a factor of great importance to probationary faculty, faculty seeking promotion to a higher rank, and other researchers and scientists for whom not merely the quantity but quality of publication weighs heavily in the evaluation process. Alternatively, publishing outside of one's field in disciplinary journals more open to collaborative research does not fully resolve the matter, regardless of how prestigious those venues might be. As Fry (2001: 163) notes, "[there are] significant differences between subjects in the merit required at each stage in the career ladder."

To illustrate this point, a sociologist may benefit greatly by publishing in *The Bulletin of the American Meteorological Society*, but meteorology faculty, depending on the institutional values and norms of the University review committees, may have less to gain from publishing in the *American Journal of Sociology*. An alternative strategy might involve all parties electing to publish in an interdisciplinary journal, but doing so offers little more guarantee that work will be regarded as worthy in the eyes of one's peers. The problem is perhaps a circular one: faculty are hesitant to publish interdisciplinary work, interdisciplinary research accordingly does not advance and correspondingly is not regarded as valuable, and therefore faculty are hesitant to acknowledge interdisciplinary work, *ad infinitum.* Indeed, this is not merely conjecture. Barthel and Seidl (2017) developed a database of groundwater research articles from 1990 to 2014. Applying a variety of metrics, including variation in the disciplinary composition of authors, references and citations, and other factors, each journal was ranked according to degree of disciplinarity and multidisciplinary. To the point of discussion, and perhaps unsurprisingly, when comparing prestige and impact factors of the two journals, it was found that multidisciplinary journals enjoyed much less prestige in comparison to disciplinary peer-reviewed publications. This may have resulted in the small increase in the number of journals regarded as multidisciplinary over time. Again, without incentive, there may be no reason to expand interdisciplinary journal offerings.

Assessment of Success

The success of an interdisciplinary project, as well as future interdisciplinary projects, be they in the realm of meteorology or elsewhere, requires routine and comprehensive assessment. The question of assessing an interdisciplinary venture remains twofold. On the one hand, the overall aims of the project must naturally be taken into account. As with any project, traditional metrics naturally apply to interdisciplinary collaborative projects between the natural and social sciences. That is to say, metrics related to outcomes and innovation—were the technical aims of the project achieved? Was technology advanced in some way? However, other dimensions of the project, unique to the interdisciplinary venture, require appraisal moving forward. To wit, Stokols et al. (2008: 81) summarizing Croyle (2008), provide a comprehensive set of assessment goals for any interdisciplinary project that takes into account traditional benchmarks of success while also including interdisciplinary assessments, as well. Although Stokol's model deals specifically with medicine and health viz cancer treatment, the model for evaluation is general enough to be applied in a variety of other contexts. Accordingly, outcomes on an interdisciplinary project generally and comprehensively tend to include:

> (1) scientific advances in a targeted area of research (e.g., cancer communications or tobacco-use research) resulting from collaborative synergies within and across participating research centers; (2) innovative approaches to and intended outcomes of transdisciplinary research training; (3) translations of scientific research into useful and sustainable clinical practices and community health programs; (4) translations of scientific research into innovative health-policy initiate es; and, ultimately; (5) reductions in health-risk behaviors, health disparities, and the incidence of chronic diseases within a particular population. (Stokols et al. 2008: 81; summarizing Croyle 2008)

This comprehensive rubric for assessment is useful because it measures several dimensions. The natural scientific dimension is assessed—does the new technology so designed provide benefits above and beyond that which has existed in the past? Has new knowledge been developed in the form of testable theory? Has proof of concept been sufficiently established to warrant future development? Naturally, without some technical outcome, or some advancement however incremental in technology or thinking about a technical problem, the project would not be regarded as a success by any measure of the term. Yet, at the same time, metrics of interdisciplinary success are included, including the translational and health-impact outcomes. Accordingly, Stokols et al. (2008) and Croyle (2008) offer a useful way to evaluate the success of collaborative projects between the natural and social sciences.

A Blueprint for the Design of an Interdisciplinary Project

Previously in Chap. 5, we identified (a) an expanded set of roles (e.g., mentor, facilitator) for interdisciplinary research and in Chap. 6 (b) some common and potential problems that emerge within the context of an interdisciplinary effort, along with some ways to address them. Here we synthesize these ideas into an outline that generally describes how to establish and manage an interdisciplinary project involving collaboration between STEM and social science. The goal of doing so is to take vaguely conceptual instructions for developing interdisciplinary research ("involve multiple disciplines," "ensure thematically based questions," "make sure projects are research integrated," etc.) and ground them in concrete principles that scholars, researchers, and stakeholders can adapt accordingly to the needs of their own projects.

Overall, all stakeholders—from core science researchers to them most distant of beneficiary end-users— participating in an interdisciplinary project must work together to develop collaborative methodologies aimed squarely at answering the global question established from the outset. Prior to the study, a plan for collaboration must form around the global question and be maintained through routine feedback loops on the basis of which methodologies are created and refined through relevant disciplinary partnerships. We advance here a rough guide or roadmap for the development of a severe weather interdisciplinary project. What is more, the guide is general enough to be adaptable to the development of any interdisciplinary project of any theme or topic area—the only condition, once again, is that there exists some applied scientific question or questions.

Define the Problem Globally

Namely: what is the *ultimate* goal of the proposed technology? As noted, a vast majority of applied research questions, whether in the realm of engineering or applied natural science, address a problem, and the problem, while initially appearing as a technical one, at the end of the day, inevitably deals with an observed problem in society—typically issues with health, well-being, or security. In solving this problem, the technology performs the *useful function* so described by the NSF. Once again, a severe weather warning project is not necessarily about radars per se, but about radars protecting lives and property, which is an altogether different question calling for an altogether different approach. As noted earlier, perhaps the most efficient and simple way to develop an interdisciplinary project, therefore, is to identify some problem or need the resolution of which may come through scientific research.

The first step is to very broadly define a practical need or problem—for instance, the problem of severe weather mortality or severe weather socioeconomic consequences. Discussions with scientists, emergency managers, and policymakers are key here since these groups tend to understand the needs of the community. Equipped with a general sense of community needs, we might in turn deduce several key research questions around which interdisciplinary research could be designed and developed in order to address those needs. For example, assume an interdisciplinary project on hurricane shelter infrastructure is to be developed. In addition to making shelters stronger and more storm resistant, we might also consult local officials, emergency managers, weather forecasters, and members of the community to understand challenges in shelter access during the storm. Do all groups normally access the shelter? What prevents them from doing so? Will people with children or pets hesitate to seek refuge if shelters are not child or pet friendly? Again, this is all pursuant of the global question from which all interdisciplinary research must begin: are the shelters actually reducing mortality through use, or are they merely more storm resistant? If truly interdisciplinary, the project cannot simply be about improving structural integrity or wind resistance of the public hurricane shelter (although this is naturally no small part). While the technical question would be "how do we design shelters to better resist hurricane-force winds," the *global* interdisciplinary question is: "how do we *improve* shelters to save lives." Here, "improvement" has a dual meaning: to both improve the structural strength of shelters *and* encourage their use by understanding the social, psychological, and cultural barriers to shelter-seeking during a hurricane event. Researchers should, from the beginning, clearly articulate hypotheses regarding the social and natural scientific dimensions of the problem, or key issues or challenges that emerge from either the natural or social realm. Indeed, this should be regarded as a condition for beginning any interdisciplinary project: the establishment of the global question as a metric of success.

Identify Key Stakeholders Early and Incorporate Them into the Design Process

As discussed at length, another hallmark of interdisciplinary research is the presence of multiple stakeholder groups. Stakeholders are once again indispensable partners in the interdisciplinary realm at all stages of the research process. At the beginning, stakeholders should be organized into focus groups, the results of which help to establish the global question discussed earlier. The global questions cannot be established without initial input from stakeholders and end-users. These focus groups reveal, among other things, common problems in the community, potential problems with the current system (and how innovation might address those problems), and also identify other stakeholder groups who might be involved in the project. Based on this, and after generating a series of hypothesis, researchers should then deduce key end-user groups involved in the project, further subdividing stakeholders, broadly, into core, mid-range, and beneficiary end-users. Stakeholders should then, in partnership with the core researchers, take a lead role in reviewing the hypotheses. A promising format in which such a review might take place would be a focus group, the advantage of which is the potential for real-time critique and revision of core project themes. In the initial stages of the project, the following might be used as a model:

1. Identify potential stakeholders through the literature, government websites, policy manuals, and community contacts
2. Decide which end-users will be included in the core, mid-range, and beneficiary end-user groups
3. Develop focus groups with relevant stakeholder communities. Towards refining the global question, among others, questions for stakeholders should include:
   (a) What are the current limitations in the present state of system design or technology, particularly from a social perspective?
   (b) Who is largely affected by the technology?
   (c) What are common day-to-day and long-term planning activities?
   (d) What would be needed from future technology?
   (e) If we hypothetically decide to design the project in this way, what will be the advantages and disadvantages of doing so?
   (f) Other stakeholder groups who might be affected by the technology?

Develop a Budget Capable of Addressing the Global Question: Not Merely Technical Innovation

If grant funded, great thought and reflection is needed regarding the budget during the proposal process. Ideally, before grant funding is sought, the global question should be developed through initial stakeholder engagement. This might require initial seed funding or funding or other small grant support, but is altogether critical for the development of a large, long-term budget for any interdisciplinary effort. Budgeting decisions and distribution should be made with reference to the global question: funding must be sufficient to address both the demands of technical innovation, but must also be sufficient to support the research activities of social scientists whose research intends to integrate users into system design. All else equal, social science budgets will be less than that of natural science budgets. On the other hand, the development of a radar network could easily run into the millions of dollars. Bearing these figures in mind, the social science budget must remain solvent enough to fulfill the integrative purposes of the project, or one commensurate with the scope necessary to fully capture the social dynamics involved in the technical system at hand. Otherwise, an important dimension of the interdisciplinary vision—that of social integration—will remain unfulfilled. Again, we will have technical integration, but not social integration. The global question will remain unanswered.

Establish an Ongoing Culture of Interdisciplinarity

One of the most important—if not the most important—requirements of an interdisciplinary project is the establishment of a culture of interdisciplinarity from the beginning, as well as the maintenance of a cooperative, multidisciplinary culture over time. This involves the need to develop a "common" language, identifying critical concepts, and defining basic elements so that we can develop a mutual understanding of what we are aiming to achieve, why, and how. The first critical issue is that we can have conversations in which we can understand each other. Also, fostering teamwork is essential, where we identify everyone's roles and responsibilities and how they all contribute to achieving our agreed upon goals. From inception through completion, participants should be given a clear definition of the global question. Here, the project facilitator should work with project engineers and natural scientists to understand the intended technical approach of system design, and, accordingly, work with natural scientists to identify key assumptions in these approaches in order to evaluate those assumptions through research. On this basis, key partnerships should be developed. Will the sociologists work with engineers? The political scientists with meteorologists? The economists with forecasters? Partnerships should be determined based on how well the scope of each discipline "fits" with questions intended to integrate the engineering goals with the needs of those whom the technology is intended to help. Economists and civil engineers may work together on a question dealing with the financial costs of new shelter designs. Psychologists might work with computer scientists to measure how well new forms of weather data resonate with forecasters and emergency managers. Determining who will be working with whom, in what regard, and on which questions and assumptions is a critical hurdle to be crossed in the beginning stages of an interdisciplinary project. Naturally, since there are many uncertainties involved in system development, research partnerships will develop and evolve as the project moves forward, but a general framework for collaboration at project conception establishes the consistency and expectations necessary to avoid partnerships of an impromptu character. While the need to adapt and modify approaches will inevitably come about, the need to alter key scientific protocols established at the beginning of the project may be avoided though a conscious strategy for collaboration from the beginning.

How might the establishment of an interdisciplinary culture look in practice? There may be two ways in which to establish the interdisciplinary efforts of interdisciplinary teams. The planning process must at first determine workable, interdisciplinary questions around which collaborative partnerships are assigned by discipline. In so doing, a structured plan for teamwork across the discipline is built and all involved know on what and with whom the work will take place. However, this is only one part of the interdisciplinary equation. Second, space should also be given to allow informal partnerships to emerge. While the need for structure, organization, and interdisciplinary culture has been a point of emphasis, projects should not be so restrictively organized as to discourage interaction between members who may have not originally been tasked to work together. Once members of an interdisciplinary project convene, previously unthought-of partnerships and collaborations may come about, and, indeed, proximity is a primary mechanisms through which this is achieved. In a study of interdisciplinary partnerships Kraut, Egido, and Galegher (1990) examined collaborative research among scientists, particularly with respect to the role of proximity on the process and outcomes of collaborative scientific research. Surveying roughly a sample of 90 subjects from a population of 500 PhD and MS researchers and scientists in the natural sciences, social sciences, engineering, and computer sciences, the authors looked at the relationship between proximity and collaboration. Lending further credence to the notion that face-to-face, informal interaction is a key driver of interdisciplinary partnerships, the authors found that physical proximity alone had an independent effect on the likelihood of collaboration. Proximity, it was theorized, creates and sustains collaboration because "frequent, informal, and...unplanned contact provides a mechanism for both bringing together potential partners as well as for maintaining existing collaborative relationships" (166). Cell phones, zoom meetings, and emails, while making communication more efficient, cannot take the place of in-person interaction when it comes to interdisciplinary research it would appear.

Finally, as far as developing an interdisciplinary culture goes, it is once again important, if at all possible, to include among project researchers *at least some* members who have previously collaborated in an interdisciplinary setting. Cummings and Kiesler (2008) studied the effects of proximity on project success through an analysis of collaboration on Information Technology Research (ITR) Program of the U.S. National Science Foundation. The study surveyed 885 ITR senior personnel on their relationships within the project, specifically concerned with what personal, policy, and organizational-structural factors might impact the collaborative tie strength among interdisciplinary team projects. While distance between members held a very strong negative relationship with collaborative tie strength, prior collaborative experiences held a positive association with tie strength, implying that a past history of collaboration strengthens cooperation and bonds in future interdisciplinary work. The authors summarize: "These findings suggest that when assessing the forces that promote strong collaboration in interdisciplinary and distributed projects, it is important to consider the connections among all project members. If pairs of senior personnel have had no prior experience with one another, then coming from different universities and different disciplines will reduce their chances of developing strong working relationships in subsequent projects" (n.p.).

Plan to Manage Conflict and Morale Issues

Natural disagreements occur within an interdisciplinary academic setting almost by virtue of design. It is perhaps inevitable, at least to a degree, that some differences of opinion may come about in a setting wherein individuals from diverse scholarly, epistemological, and methodological backgrounds are tasked to collectively solve a problem. Even within simple monodisciplinary projects, the scholarly venture is one characterized by contentious ideas and debates. The advantage of disciplinary fields is that debates tend to occur over established ideas, established methodologies, and established theories. Interdisciplinary projects, on the other hand, contend with an additional layer of potential dispute, that of potentially incompatible ontologies and epistemologies between disciplines with drastically different methods, theories, and approaches. In sociology, for instance, no parallel methodology exists to the collection and analysis of weather data; similarly, no similar method exists in meteorology to that of qualitative fieldwork, which is a common approach in sociology. Thus, social scientists and natural scientists "see" the world in different terms, and, as a consequence, regard certain questions as important or unimportant; modes of inquiry as relevant or irrelevant; and theories as either applicable or inapplicable. The main point of conflict in interdisciplinary studies occurs at a point where a question emerges that overlaps both the domain of inquiry established in both the social and natural sciences, thereby creating under some circumstances the potential for conflict.

Solving these challenges remains critical because evidence suggests that an inattention to the unique needs of the interdisciplinary effort has led in some cases to morale problems among participants. In this respect, Kraut, Egido, and Galegher (1990) found that psychology faculty reported far less enthusiasm for the process of working collaboratively on research projects. Nearly all of the publications on which respondents collaborated with co-authors yielded more negative attitudes on questions related to satisfaction with the research process. According to results, when undertaking collaborative research, enthusiasm declined most significantly in relation to the perceived importance of work, and managerial tasks associated with the project were consistently and significantly reported as more difficult when working with colleagues or associates. Satisfaction with the research process, as well as the apparent ease of scientific and technical tasks, also ranked lower when performed in a collaborative environment. The outcomes suggest, naturally all else equal, that researchers seem to find collaborative research far less rewarding than research conducted singlehandedly, perhaps owing to the fact that interdisciplinary research raises unique logistical and interpersonal difficulties less common among purely disciplinary pursuits.

Conclusion

Over the past five decades, there has been an increasing tendency towards collaboration across the disciplines in the conduct of scientific research (Wuchty et al. 2007). This is perhaps not unexpected. Given the increasing growth and complexity of society, and, accordingly, the growing complexity of the problems with which society is confronted, it is only natural to assume that such tendencies towards interdisciplinarity will likely increase. Indeed, collaborative research has become a regular dimension—if not defining feature—of research in a variety of fields, including, but not limited to, meteorology (Goodsite and Gammelgaard 2016), landscape agriculture (Fry 2001), medicine (Albert et al. 2015), and even software engineering (Lenberg et al. 2015). There is indeed no shortage of collaborative research at present nor will there likely be in the future. It is therefore perhaps not unreasonable to conclude that the future will see interdisciplinary research as the rule, rather than the exception, particularly in the context of applied research where both nature and society play interconnected roles. This will call for an increased focus on developing interdisciplinary models for collaboration, of which there are presently few. The preceding two chapters have been an attempt to move the field in this direction, but to move forward in the development of interdisciplinary research, severe-weather-related or otherwise, additional empirical work is necessary

In Chap. 7, we turn to a case study through which many of the concepts and ideas discussed in the present chapter are illustrated. The Engineering Research Center for the Collaborative Adaptive Sensing of the Atmosphere (ERC-CASA), an interdisciplinary severe weather warning project with which the authors were involved, from roughly 2000–2010, will serve as a basis for these illustrations.

Bibliography

Albert, M., Paradis, E., and Kuper, A. (2015). Interdisciplinary Promises versus Practices in Medicine: The Decoupled Experiences of Social Sciences and Humanities Scholars. *Social Science & Medicine*, 126, 17–25. doi: https://doi.org/10.1016/j.socscimed.2014.12.004.

Barthel, R and Seidl R. (2017). Interdisciplinary Collaboration between Natural and Social Sciences - Status and Trends Exemplified in Groundwater Research. *PLoS One*, 12(1), e0170754.

Boaz, A., et al. (2018). How to engage stakeholders in research: design principles to support improvement. *Health Research Policy and Systems*, 16, 60.

Bracken, L. J., and Oughton, A.E.. (2006). What Do You Mean?' The Importance of Language in Developing Interdisciplinary Research. *Transactions of the Institute of British Geographers*, 31(3), 371–82. doi: https://doi.org/10.1111/j.1475-5661.2006.00218.x.

Catney, P. and Lerner, D.N. (2009). Managing Multidisciplinarity: Lessons from SUBR:IM. *Interdisciplinary Science Reviews*, 34(4), 290–308. doi: https://doi.org/10.1179/030801809X12529269201129.

Committee on Facilitating Interdisciplinary Research, Committee on Science, Engineering, and Public Policy (2004). *Facilitating interdisciplinary research*. National Academies. Washington: National Academy Press, p. 2.

Croyle, R.T. (2008). The National Cancer Institute's transdisciplinary centers initiatives and the need for building a science of team science *American Journal of Preventative Medicine*, 35(2S), S90–S93.

Cummings, J.N., and Kiesler, S. (2008). Who Collaborates Successfully? Prior Experience Reduces Collaboration Barriers in Distributed Interdisciplinary Research. Pp. 437–46 in *Proceedings of the 2008 ACM conference on Computer supported cooperative work, CSCW '08. New York, NY, USA: Association* for Computing Machinery.

Cummings, J.N., and Kiesler, S. (2005). Collaborative Research Across Disciplinary and Organizational Boundaries. *Social Studies of Science,* 35(5), 703–722. doi:https://doi.org/10.1177/0306312705055535

Deverka, P. A., Lavallee, D. C., Desai, P. J., Esmail, L. C., Ramsey, S. D., Veenstra, D. L., & Tunis, S. R. (2012). Stakeholder participation in comparative effectiveness research: defining a framework for effective engagement. *Journal of comparative effectiveness research*, *1*(2), 181–194.

Fischer, A. Tobi, H. and Ronteltap, A. (2011). When Natural met Social: A Review of Collaboration between the Natural and Social Sciences. Interdisciplinary Science Reviews, 36(4), 341–358.

Fitzgerald, C., Martin, A., Berner, D. *et al.* (2019). Interventions designed to reduce implicit prejudices and implicit stereotypes in real world contexts: a systematic review. *BMC Psychol,* 7, 29.

Fry, G. (2001). Multifunctional landscapes—towards transdisciplinary research. *Landscape and Urban Planning*, 57(3–4), 159–168.

Goodsite, M. and Gammelgaard, A. (2016). Interdisciplinary Research in Meteorological Sciences. *Journal of Climatology & Weather Forecasting*, 1(1), doi: https://doi.org/10.4172/2332-2594.1000e101.

Grimshaw, J.M., Eccles, M.P., and Lavis, J.N. *et al.* (2012), Knowledge translation of research findings. *Implementation Science,* 7, 50.

Haapasaari, P., Kulmala, S., and Kuikka, S. (2012). Growing into Interdisciplinarity: How to Converge Biology, Economics, and Social Science in Fisheries Research? *Ecology and Society, 17*(1), 6.

Hagstrom, W.O. (1964). "Traditional and Modern Forms of Scientific Teamwork." *Administrative Science Quarterly*, 9(3), 241–263. doi: https://doi.org/10.2307/2391440.

Haney, A. and Power, R.L., (1996). Adaptive management for sound ecosystem management. *Environment Management*, 20, 879–886.

Harris, F. and Lyon, F. (2013). "Transdisciplinary Environmental Research: Building Trust across Professional Cultures." *Environmental Science and Policy,* 31, 109–19. doi: https://doi.org/10.1016/j.envsci.2013.02.006.

Jeffrey P. (2003). Smoothing the Waters: Observations on the Process of Cross-Disciplinary Research Collaboration. *Social Studies of Science*, 33(4):539–562.

Kirby CK, Jaimes P, Lorenz-Reaves AR, and Libarkin JC (2019). Development of a measure to evaluate competence perceptions of natural and social science. *PLoS ONE*, 14(1), e0209311.

Klein, J. T. (1990). *Interdisciplinarity: History, theory, and practice.* Wayne State University Press.

König, B., Diehl, K., Tscherning, K. and Helming, K. 2013b. "A Framework for Structuring Interdisciplinary Research Management." *Research Policy* 42(1):261–72. doi: https://doi.org/10.1016/j.respol.2012.05.006.

Kohlhagen, T., Fryirs, K. and Semple, A.-L. (2013). Adaptive management in upland swamps. *Geographical Research*, 51(4), 439–453.

Kraut, R. E., Egido, C., & Galegher, J. (1990). Patterns of contact and communication in scientific research collaboration. In J. Galegher, R. E. Kraut, & C. Egido (Eds.), *Intellectual teamwork: Social and technological foundations of cooperative work* (pp. 149–171). Lawrence Erlbaum Associates, Inc

Lélé, S. and Norgaard, R.B.. (2005). "Practicing Interdisciplinarity." *BioScience*, 55(11), 967–975. doi: https://doi.org/10.1641/0006-3568(2005)055[0967:PI]2.0.CO;2.

Lewontin, R. and Levins, R. (1998). How Different Are Natural and Social Science? *Capitalism Nature Socialism*, 9(1), 85–89. doi: https://doi.org/10.1080/10455759809358780.

Machlup, F. (1988). Are the social sciences really inferior? *Society*, 25, 57–65.

McFadden, L., Penning-Rowsell, E., and Tapsell, S. (2009). Strategic Coastal Flood-Risk Management in Practice: Actors' Perspectives on the Integration of Flood Risk Management in London and the Thames Estuary. *Ocean and Coastal Management*, 52(12), 636–45. doi: https://doi.org/10.1016/j.ocecoaman.2009.10.001.

Mitroff, I., and Emshoff, J. (1979). On Strategic Assumption-Making: A Dialectical Approach to Policy and Planning. *The Academy of Management Review*, *4*(1), 1–12.

Nuijten, E. 2011. "Combining Research Styles of the Natural and Social Sciences in Agricultural Research." *NJAS: Wageningen Journal of Life Sciences* 57(3–4):197–205. doi: https://doi.org/10.1016/j.njas.2010.10.003.

Lenberg, P., Feldt, R. and Wallgren, L.G.. (2015). Behavioral software engineering: A definition and systematic literature review. *Journal of Systems and Software*, 107, 15–37.

Quinn, R.E. (1988). *Beyond Rational Management: Mastering the Paradoxes and Competing Demands of High Performance*. San-Francisco, CA: Jossey-Bass

Rhoten D., O'Connor E., and Hackett E.J.. (2009). The Act of Collaborative Creation and the Art of Integrative Creativity: Originality, Disciplinarity and Interdisciplinarity. *Thesis Eleven*, 96(1), 83–108.

Roy, E.D., Morzillo, A., Seijo, F., Reddy, S. and Rhemtulla, J.M., Milder, J.C., Kuemmerle, T., and Martin, S. (2013). The Elusive Pursuit of Interdisciplinarity at the Human—Environment Interface. *BioScience*, 63(9), 745–53. doi: https://doi.org/10.1525/bio.2013.63.9.10.

Stokols, D., Hall, K.L. Taylor, B.K., and Moser, R.P. (2008). The Science of Team Science: Overview of the Field and Introduction to the Supplement. *American Journal of Preventive Medicine*, 35(2), S77–S89.

Hampton, S.E. and Parker, J.N. (2011). Collaboration and Productivity in Scientific Synthesis, *BioScience*, Volume 61(11), 900–910.

Wuchty, S., Jones, B.F., and Uzzi, B. (2007). "The Increasing Dominance of Teams in Production of Knowledge." *Science*, 316(5827), 1036–39. doi: https://doi.org/10.1126/science.1136099.

Chapter 7
A Case Study in Interdisciplinarity: CASA (Collaborative Adaptive Sensing of the Atmosphere)

The previous Chaps. 5 and 6—outlined a basic model for interdisciplinary collaboration. Namely, identifying a set of organizational roles, rules, and guidance for a long-term, institutionalized interdisciplinary research project. In so doing, the chapters provided a general "roadmap" of sorts, moving from the inception of a project through conclusion, grounded insofar as possible in the empirical literature for developing scientific and end-user partnerships across the disciplines. In this chapter, we turn our attention to that of an extended case study, an analysis and description of which will serve to illustrate more specifically the common elements of interdisciplinary collaboration between the natural (engineering included) and social sciences within the area of interdisciplinary meteorology. The project, ERC-CASA, (Collaborative Adaptive Sensing of the Atmosphere; henceforth simply "CASA"), which brought together a diverse team of scholars and stakeholders to research remote sensing technologies, will serve as the basis for the forthcoming analysis. At the time of inception, roughly the early 2000s, CASA represented what was, in reality, the first large-scale collaborative effort to develop warning systems in partnerships between STEM and the social sciences. Headquartered at UMass-Amherst, included in the project were a broad range of participants, spanning in scope an equally broad range of disciplines, stakeholders, and practitioner groups, including, but not limited to

- Engineers
- Computer Scientists
- Meteorologists
- Hydrologists
- Sociologists
- Human factors analysts
- Political scientists
- Administrative personnel
- Industrial collaborators
- NWS Forecasters

W. Donner et al., *Severe Weather Warnings: An Interdisciplinary Approach*,
https://doi.org/10.1007/978-3-031-05031-2_7

- Private forecasters
- Emergency managers
- General Public

The collaborative effort thus brought together a variety of individuals from a variety of different backgrounds and skillsets, many of whom found themselves in a position of working with scholars and researchers from different disciplines for the first time.

ERC-CASA: A Brief Overview[1]

Conceived as a partnership between academia and industry, ERC-CASA formed part of the 2nd generation of NSF ERCs, the purpose of which was in part to better and more fully capture the need for integrated systems in applied scientific research. CASA specifically brought together engineers, natural scientists, forecasters, and social scientists to develop, through improvements to radar technologies, more accurate and reliable warnings systems above what was available at the time, the ultimate goal of which was to reduce fatalities and injuries due to tornadoes and flash floods (what once again might be referred to as the "global question" discussed earlier).

Radars of the day were limited in a variety of ways, the recognition of which drove ERC-CASA into being. In the simplest sense, radars do not provide photorealistic images of storms, as if one were looking at a picture, but rather employ radio waves to return an abstract, graphical representation of storm location, structure, density, and evolution. The origins of ERC-CASA lay mainly with longstanding limitations in what had been up to that point the prevailing technological and operational paradigms of radar use for severe storms: large, powerful WSR 88-D doppler radars able to return robust data visualizations, but which were nevertheless slow to update and oftentimes of limited utility when scanning the lower atmosphere (Chandrasekar et al. 2008). An effect of radar beams traveling at a linear trajectory in space, forecasters are left somewhat blind to the lower levels of storm activity (Storm Prediction Center: https://www.spc.noaa.gov/faq/tornado/doppler.htm). In other words, since a radar beam cannot be curved to circumnavigate the horizon, radars are unable to "see" potential storm activity beyond a certain range. This is particularly challenging for the detection of tornados, indeed one present since the very inception of radar technology, given that the most discernable

[1] It should be strongly emphasized to the reader that here we do not attempt to fully describe every aspect of the ERC-CASA—it is far to large and complex to fully capture in a fully book, let alone a chapter. Rather, in the spirt of this volume, we attempt to focus on and describe, to the extent possible, the integration of social science data with the design decisions. Nor is there an attempt to describe in full detail every element and finding of the CASA project. An attempt is made to simply discuss and articulate the core elements of what made ERC-CASA interdisciplinary in an attempt to better illustrate the concepts presented in earlier chapters.

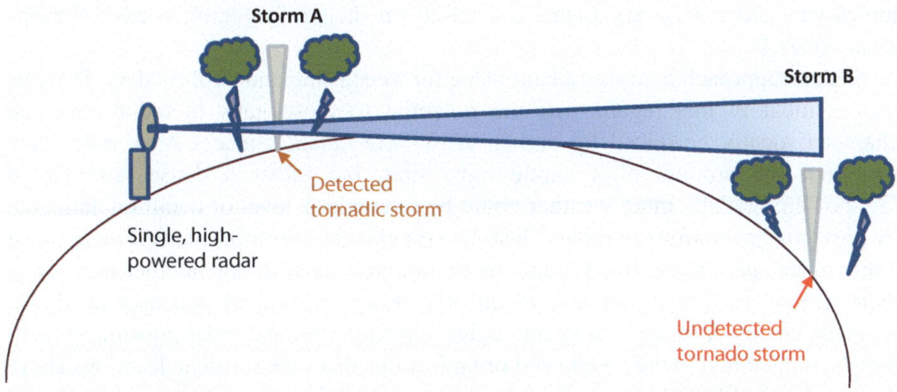

Fig. 7.1 Single Doppler radar. Note in the above illustration that the radar is limited by the curvature of the earth. The beam cannot detect the farther storm because the beam is blocked by the earth's horizon. Thus, while one of the storms is detected, the other is undetected. This has been an historical limitation in traditional Doppler set ups

severe weather pattern signatures (e.g., "hook echoes") tend to form at lower levels in the lower atmosphere. Since the reliability and accuracy of warnings are themselves limited by the reliability and accuracy of radar data, innovation in the area of remote sensing was seen as a way to improve warnings, and, as a result, improve response and safety to a range of weather-related hazards.

To do so, however, would require both improvements in the technology of radar systems as well as a fundamental change in the basic philosophy of remote sensing in order to address flaws inherent to traditional radar design and application. From a technical standpoint, this was the central problem that gave rise to ERC-CASA.

Through innovation, ERC-CASA represented a great leap forward in remote sensing theory in seeking to address what had historically been the most fundamental limitations of radar design. Prior to CASA, radar development focused principally on improving the power of systems, paying little attention to the aforementioned problems created by the curvature of the earth and other obstructions such as building, mountains, and other features of the natural and built environment. To remedy the dilemma, CASA favored developing a dense and broadly-distributed network of low-powered radar units able to "see past" common obstructions in the environment, most notably blind spots created by the curvature of the earth (Brotzge et al. 2006, 2010; Chandrasekar et al. 2007). Although the proposed radar units were to be far less powerful than the traditional WSR 88-D unit, this was made up for in terms of geographic scale: a greater number of units could be distributed more broadly and densely across the earth's surface. Whereas one large, powerful radar could easily miss an event, when dozens of small units perform scanning tasks simultaneously, an event going undetected is far less likely (See Fig. 7.1). Christened "DCAS," or "Distributed Adaptive Collaborative Sensing," "these systems map wind, rain, and thermodynamic variables in the lower troposphere, supplying real-time, dynamic data to decision makers, such as National Weather Service (NWS) forecasters and emergency managers, based on their information needs" (Philips et al. 2007: 1).

Such an approach held significant value for forecasting and public safety. Perhaps most critical in this regard was the potential to close gaps in radar coverage characteristically produced by earlier WSR-88D radars. Since CASA radar data were not the product of a single radar unit, but rather a dense network of low-powered radars, more weather could be seen with a level of detail unattainable by previous generations of radar. "Installed on existing rooftops and cell towers just a few miles apart, these small radars will communicate with one another and adjust their sensing modes in response to quickly changing weather and user needs—a dramatic change from current technologies. Up-to-the-second radar information will then be transmitted to the people and organizations that make critical decisions about the weather." (ERC-CASA n.d.-a). Within the DCAS system, if one radar's beam was obstructed by, for instance, a building, several other radars in the network, by virtue of being located in different geographic positions, could make up for this loss of vision. What furthermore distinguished DCAS from older technologies included radars that rotated more rapidly. This allows not only for better coverage, but also quicker storm updates and better tracking of storm evolution and movement. Storms that could be detected earlier, and whose movement was more precisely known, could be warned more quickly and with fewer false alarms and missed events.

For the purposes of discussion, and the relevance for the interdisciplinary approach with which this volume is concerned, there are perhaps two important benefits regarding CASA worth discussing at some length before moving forward. First, under the DCAS innovations so described, fewer severe storms will go undetected on average relative to WSR-88D technology. As well, storm scans within a given DCAS network will return data of a much higher quality and a much higher level of detail, both of which are key to determining storm structures, which is, in turn, is key to identifying data patterns indicative of severe weather within a storm system. Ultimately, this would lead to timelier and more accurate severe weather warnings. Here perhaps a further illustration of this matter is in order. Under Fig. 7.1, which represents a single WSR-88D scanning scenario, *Storm A* is detected by the single remote sensing unit because the beam is unrestricted by the curvature of the earth or any built or natural obstructions. *Storm B*, however, would go undetected because the horizon obstructs the line of sight through which the radar might detect the storm and therefore a tornado should one spawn. Turning attention to Fig. 7.2, which illustrates the DCAS system under a similar scenario, note that the system of distributed radars successfully detects *Storm A* (as did the single radar in the first illustration). However, unlike the first illustration, note that under DCAS *Storm B* the distributed nature of the system now overcomes the problem associated with the curvature of the earth blocking line of sight. Thus, with ERC-CASA's radar system, storms are more reliably observed leading to fewer undetected tornadoes, and, therefore, more tornadoes accompanied by warnings.

Second, under DCAS, rapid and detailed updates of storm data were made possible, allowing forecasters to better observe the formation and location of storm systems, making tornadoes detectable and predictable under normal conditions. Note

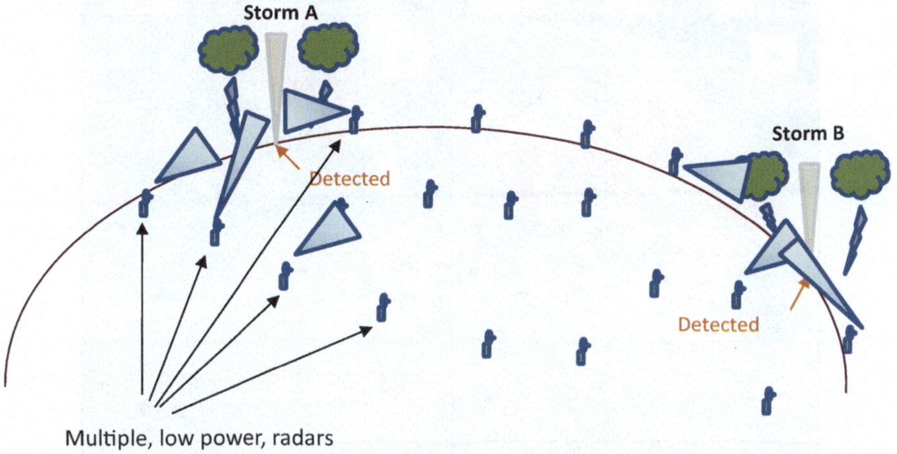

Fig. 7.2 DCAS network. A key feature of the system is that it requires less power and is of lower cost. Note how the distributed nature of the network makes it easier to accommodate curvature. For a fuller description of the DCAS system, see Zink et al. (2010) and Philips et al. (2007) for a more in-depth description of the DCAS approach to remote sensing

in Fig. 7.3 the much higher level of detail under (a) and (b) (DCAS data) compared to the far more granular visualization presented in (c) and (d) (traditional Doppler data). This innovation in graphical fidelity would allow forecasters to more easily identify data structures (e.g. hook echoes) that suggest the presence and evolution of severe weather, leading to fewer false positives and negatives in forecaster evaluation of severe weather (recall that the final forecasting judgment comes down to human judgement on the part of the forecaster). DCAS thus offers a means to identify and predict storms in the short term, which was expected to translate into better public safety through, for instance, longer lead times and fewer false alarms. However, the question remains: what is the mechanism or mechanisms by which these technological improvements—however impressive or revolutionary on their own—are translated into fewer fatalities, fewer economic costs, and fewer community impacts. Could the explanation be reduced simply to shorter lead times or false alarms, or are other factors at play? This is the "global question" of the project, one that cannot be answered through technical innovation alone. Here, in this chapter, we tell the story of how this global question emerged and evolved throughout the project.

Organization of CASA

Housed centrally at University of Massachusetts-Amherst, ERC-CASA brought together a wide consortium of universities, research centers, community stakeholders, industrial partners, and private research organizations as part of an

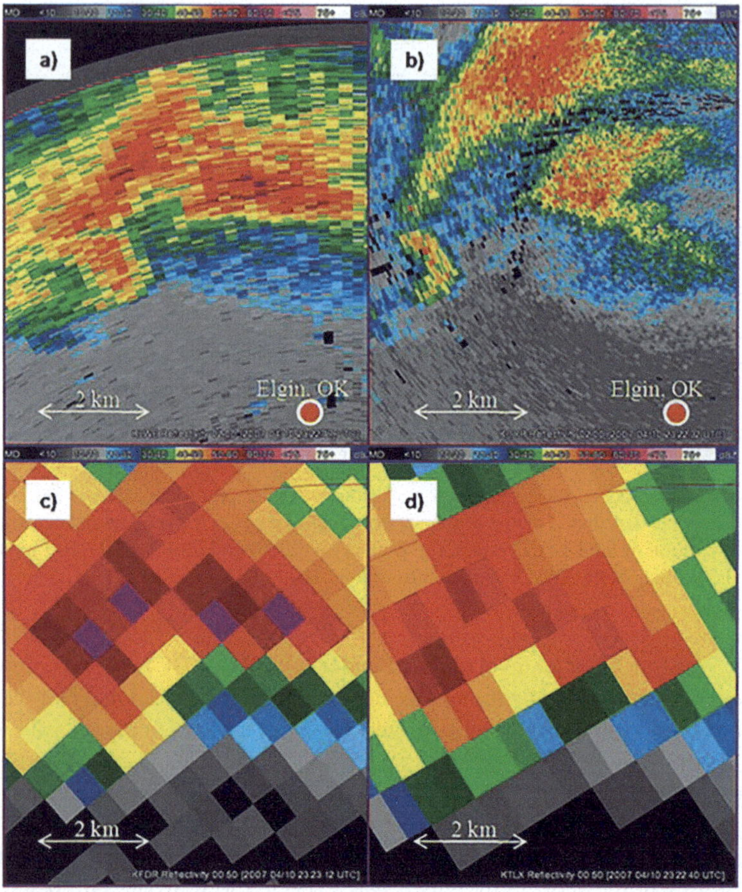

Fig. 7.3 CASA radar resolution may be seen under (**a**) and (**b**), whereas traditional Doppler resolution may be seen under (**c**) and (**d**). Note that (**a**) and (**b**) provide much more detailed resolutions in comparison to (**c**) and (**d**), which are far more granular. Courtesy: Brotzge et al. (2010). © American Meteorological Society. Used with permission

interdisciplinary mission to advance the understanding and application of remote sensing technologies. What is notable about ERC-CASA is that it stands as perhaps one of the more complex and longstanding interdisciplinary efforts of the early twenty-first century and arguably remains so even as of today. As far as the organization of the project goes, research teams were assembled into what at the time were referred to simply as "thrusts" in the common vernacular of the project. The so-called thrusts included "sensing," "distribution," and "detection," each of which focused on separate elements of a functional radar system, and around which the basic scientific work of ERC-CASA revolved and took place. While ERC-CASA established a number of test-bed sites in which to develop proposed technology, here we focus mainly on the Oklahoma test bed since this was the site at which a vast majority of the interdisciplinary effort took place.

The *detection* team was composed principally of radar engineers and computer scientists. In the most general of terms, the principal aim of the detection thrust was to improve the rate at which tornadoes were identified as well as the rate at which updates regarding storm evolution could be developed. This was done primarily by developing the radar system through hardware and software improvements. The *distribution* thrust dealt with the fundamentally new challenges of a distributed, DCAS-driven model of remote sensing by creating the computational architecture through which to process data with the purpose of optimizing system performance. Finally, the *prediction* thrust was responsible for developing new algorithms and models through which to identify severe weather and severe weather patterns. To be successfully applied under the DCAS framework, older meteorological models were in need of revision in order to accommodate the novel data the proposed radar system was expected to produce. Working together, these thrusts formed the core scientific effort of the project whereby direct, technical improvements in the pre-existing remote sensing technology would be made.

The project in turn included an "end-user integration" component, the goals of which were to

1. Advise DCAS Adaptive Scanning Strategy
2. [Study] User Interaction with New Technology for Decision Making
3. [Explore the] Impact of DCAS Technology on Communication, Public Response, and Social and Economic Vulnerability[, and]
4. Establish the Social and Economic Value of DCAS Information (ERC-CASA n.d.-b).

With respect to ERC-CASA, and in the most general of terms, the end-user integration component aimed to build into the system a set of social scientifically-derived policies through which to guide the design of the *detection*, *distribution*, and *prediction* sub-systems (see Fig. 7.4). In so doing, the goal of system integration would be achieved through the adherence to principles of interdisciplinary design. The team included a variety of scholars from a variety of disciplines—including four main disciplinary groups: sociologists, political scientists, decision scientists, and human factors engineers. Mirroring the diversity of participants, the methodological approach was accordingly diverse, including surveys of emergency managers, emergency management simulations, surveys of the public, and end-user policy development research.

There were four main end-user research streams dealing with key stakeholder integration into system design. These included:

- **Emergency manager surveys**—the surveys intended to understand the needs of emergency mangers within the Oklahoma test bed, the major testbed area for the CASA project wherein the DCAS system was evaluated in the context of tornado severe weather.

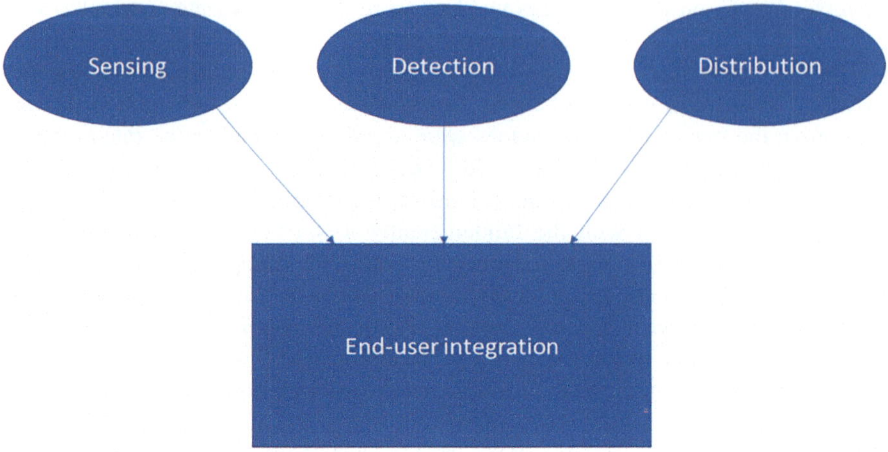

Fig. 7.4 End-user integration flow

- **End-user simulation**—conducted in part to determine how end-users—mainly emergency managers and forecasters—were observed to interact with old and new radar data.
- **Policy development**—conducted to determine how the DCAS system was to be used in practice, mainly dealing with questions about the control and use of the system.
- **Public surveys**—fieldwork conducted with the general public to understand how the public responds to tornado warnings.

Emergency Manager Surveys

As key stakeholders, Oklahoman emergency managers, as likely users of CASA's new radars, participated in a series of surveys and interviews in order to make the system more relevant and adaptable to the needs of users. The effort addressed several key themes and questions. Critically, what are some key challenges emergency managers presently faced in warning the public, could present design approaches address these needs, and, if not, how might the design process be altered to do so? What were the perspectives among interviewed emergency managers on the new system and how it might add value to the warning process? How did they rank and evaluate severe weather risks in their communities? What kinds of weather data did emergency managers commonly use and feel comfortable using, and what were some difficulties in accessing and interpreting data up to that point? Addressing these questions was an important condition for the development of DCAS since, as is recalled from Chap. 4, emergency managers, meteorologists, and forecasters make decisions in a way that is governed by social, cultural, economic, psychological, and political forces that should be considered in the design process. ERC-CASA, therefore, would need to take these factors into account in order to design DCAS in a way that was both useful and relevant to the context of severe weather warnings in Oklahoma.

Studies of Oklahoman Emergency Managers

In any interdisciplinary project, the first and perhaps most critical task involves the identification and profiling of the various stakeholder or end-user groups for whom the project is relevant in some regard. Emergency managers, as potential users of a new radar system, naturally constituted one such stakeholder group. Integrating emergency management needs into system design was vital since emergency managers are directly responsible for decisions that directly affect public safety, acting as a sort of "middleman," as it were, between the forecaster and the community. Improving warnings was central to the ERC-CASA mission, and warnings stand as perhaps the most important tool available to emergency managers when responding to tornadic storms.

The initial interviews with emergency managers yielded several interesting and unexpected insights regarding decision making, the use of radar data, and perceived challenges to public response. First, what was highly notable was that the emergency management community in Oklahoma was far from uniform, characterized, on the other hand, by a great deal of diversity on a number of different variables, particularly with respect to training, experience, and access to technology (Rodriguez et al. 2010). Among the roughly 50 emergency managers surveyed across 26 Oklahoma counties, a wide range of themes, ideas, and issues relevant to the design of CASA's radar system and policy emerged. From this, here are identified three factors that might shape and limit how emergency managers used the potential system: (1) *An emergency manager's background,* (2) *The level of data complexity, and* (3) *Social and psychological factors shaping the understanding and use of weather information.*

1. *Emergency Manager Background*

   From the beginning of the project, it quickly became evident that the emergency management community in Oklahoma was characterized by a great deal of unanticipated professional, social, and demographic diversity. It appeared no two emergency managers were alike. Some were well trained, enjoyed access to sophisticated technologies, and were well funded, heading the kind of well equipped, state-of-the-art Emergency Operations Centers (EOCs) one finds commonly portrayed in the media (Mitchell et al. 2000). A considerable number of emergency managers in rural areas, many of whom simply volunteered for the position, had little training, and operated on shoestring budgets. Most relevant to CASA, it was discovered that local emergency managers in smaller and more remote communities simply lacked access to the kinds of technology and training that would allow them to warn and protect communities as effectively as their urban counterparts (it should be noted here, however, that OK-FIRST, a program launched in 1996 to provide support to Oklahoma emergency managers, made great strides in training emergency managers in radar and data interpretation. Many communities benefitted—and continue to benefit—from OK-FIRST). These divides were not without consequence as far as warnings are concerned. Most notably, rural emergency managers consistently reported difficulties in accessing and making full use of severe weather radar data on the basis of which the decision to warn was oftentimes made (Marks 2006). This was in part due to generally slower internet connections commonly seen in rural areas due to infrastructure limitations, a pattern which continues to this day (Lai and Widmar 2021).

Technological limitations may have made CASA much less useful for some stakeholders. Many rural emergency managers at the time still relied heavily, if not exclusively, on dial-up connections (if indeed a computer was even available for use, which was not always the case). In the context of severe weather warnings, dial-up connections, due to bandwidth limitations, significantly curb the speed with which data on severe weather are updated to the individual user. Tornadoes and flash floods, particularly due to the rapidity and uncertainty with which they form and evolve, require swift updates of the kind that might be impossible under the limitation of dial up internet. Dial-up—or the absence of internet altogether—therefore naturally places rural emergency managers at a significant disadvantage as changes in the location, timing, and scope of a severe weather event may occur at a pace that far exceeds the capacity of slower connections to promptly process and visualize the relevant radar data on the event of interest. Even if access to technology were not an issue, Marks (2006) reports a second and related problem: that of training. Many rural emergency managers, by virtue of being volunteers, lacked the skills to interpret radar data of the detail and complexity the DCAS system intended to produce. In all, the information collected from emergency managers suggested, at the time, that the CASA system would be of full benefit only to a limited cross section of the emergency management community: trained, well-equipped, urban emergency managers and officials.

This information was deeply relevant for the design of ERC-CASA's DCAS network and provides here a clear and instructive example of the integrative nature of interdisciplinary collaboration. DCAS, from a purely technical perspective, was, as a consequence of design, a highly complex and complicated system, and, as a result, was anticipated to make heavy demands as far as hardware and broadband use was concerned—older computers or even newer computers lacking broadband connectivity may have had trouble accessing the DCAS network data or making effective use of resulting data, which was, by design, of a far more demanding nature. Thus, although DCAS data may have indeed held certain technical advantages over the previous generation of radar infrastructure, in a real world setting where mid-range end users differ as far as technology access and training goes, such improvements may have been of limited advantage to emergency managers without the resources to make full use of them.

2. *Data complexity*

   The kinds of information and data emergency managers require to effectively warn communities presented an important theme as well. An important question, therefore, was: what kinds of weather information do emergency managers prefer, and, furthermore, what shapes and determines these preferences? To this end, Donner et al. (2005) examined data needs and data preferences within the Oklahoma emergency manager and floodplain manager community to shed light on these preferences. Much of the discussions centered on whether interviewees would prefer better data on impending tornadic or flood threats. Emergency managers tended to favor data that would help them better predict tornadoes since floods were regarded as generally well controlled and tornadoes, on the other hand, far more unpredictable and dangerous. More easily mitigated, flooding was therefore seen as a secondary worry as far as radar data is concerned. But differences in the physical threats presented by the two events were not the sole reason for this decision. Consideration was also given to the politics of severe weather decision making, as well. Emergency managers were also aware of the profound consequences of missed tornadoes: failing to issue a warning for a tornado came with anticipated costs in public confidence and accountability that were higher in comparison to failing to warn a flood. Thus, these political costs could be avoided or mitigated with radar data better suited to tornado prediction and warning. This served as yet another reason for the preference for better tornado prediction capabilities in the proposed system.

   These findings were critical because they established a context for system design, which is the ultimate strength of the interdisciplinary approach. Knowledge of whether radar data or flood data would be more useful to emergency managers was unknown, and end-user research provided a great degree of clarification on this matter. Thus, with this knowledge in hand, the system could be designed in such a way that reflected the realities within which emergency managers made decisions rather than the untested assumptions of core scientists.

3. *Patterns of understanding and use of weather information.*

   Finally, psychological, and social factors were found to shape how emergency management end users interact with and make use of severe weather information with which they are presented. League et al. (2010) surveyed 62 emergency managers to develop an understanding of severe weather data use and application during tornadoes, floods, and other hydrometeorological threats. Using a mixed methods approach, data were collected via interviews and focus groups with emergency managers at the 2009 annual Oklahoma Emergency Management Association (OEMA) conference in order to evaluate the most common tools emergency managers use on a day-to-day basis to monitor the weather and make decisions when the possibility of severe weather presents itself. According to results, during periods of heightened risk to hazardous storms, emergency managers appeared to make heaviest use of WSR-88D radar data, NWS text messages, and spotter reports, whereas private weather information ranked among the least useful. Important here, rather than relying on a single source of weather information, findings would seem to suggest that emergency managers synthesize

> multiple weather products in order to produce a general picture of the present weather situation (although a preference for public (e.g., NWS), rather than private, sources of weather information appeared to prevail overall). What was further notable is that "raw" radar was generally avoided, which likely resulted from a general lack of expertise in interpreting radar imagery on the part of many emergency managers. In other words, emergency managers preferred "value-added" radar data—radar data to which interpretations were added by experts. To that point, League et al. found a diversity of experience and training with radar technologies, offering the recommendation that EMs be routinely trained in the use of weather information, again hearkening back to the idea that emergency managers are a diverse stakeholder group and therefore require a diversity of data and support to successfully perform.

Here there is an important lesson regarding interdisciplinary research—once again, that most applied research questions, irrespective of topic, and always naturally by design, make unavoidable assumptions about how end-users think and behave. If "built" into the system untested, and at variance with how actual users think and behave, these assumptions will limit the overall effectiveness of resulting technical innovation—again, no matter how "better" the system is from the standpoint of pure technical innovation. Absent a detailed analysis of emergency managers, "built in" to the system would have been an end-user design component with the mistaken assumptions that a majority of emergency managers share similar backgrounds, locations, training, and resources, realities we now know to be untrue. With these insights, ERC-CASA could reshape system policy to address more broadly and comprehensively the needs of emergency management, the needs of which were determined to be highly varied. Recommendations were made to accordingly reduce complexity in data insofar as possible for the untrained user, minimize hardware or software requirements as much as allowed, and provide additional training through OK-First or elsewhere with as few impacts to the sophistication of the system as possible.

End-User Simulations

Intended to assess the potential impact of DCAS on decision making, CASA human factors analysts developed a series of simulated severe weather experiments in which to observe decision making patterns among forecasters and emergency managers. The simulations sought to determine how severe weather data were used, which kinds of data were preferred, and the common challenges these two groups face when the need to detect and predict potentially deadly storms arise in an operational setting. Recall that prior to ERC-CASA, radar technology of the previous generation suffered significant design limitations, which included poor resolution, blind spots in the scanned environment, and attenuation at long distances, all of which CASA intended to address. In spite these weaknesses, there was, however, one distinct "advantage" to traditional WSR 88-D radar: resulting data were, in relative terms, much more simple and therefore interpreted with greater ease. Thus, one needed comparably less (though still considerable) education training, and, importantly, practice to make general sense of WSR 88-D radar imagery. All else equal, with previous radar generations, a novice user could be instructed in the interpretation of radar weather data to improve situational awareness of potentially-tornadic weather patterns and, in direct consequence, of warning decision making.

So to speak, DCAS presented a double-edged sword in this regard: although DCAS would yield more accurate and reliable radar data, a natural outcome of such improvements was a level of complexity significantly higher than that witnessed in data produced from WSR-88D radars. With this in mind, the interdisciplinary question then became: would the far more complex DCAS be usable by the majority of Oklahoman emergency managers, or would the data simply add to the complexity of decision making, perhaps ironically adding to the difficulty of fully assessing the weather? As noted in Chap. 4, as well as in League et al. (2010), forecasters and emergency managers make decisions on a the basis of multiple sources of weather information. Consequently, the problem here is that adding an additional, highly-complex data source to the common collection of weather data sources emergency managers use risked creating rather than reducing additional uncertainty. Given that DCAS was unprecedented, no answers to this question were immediately forthcoming.

To address the issue, weather radar simulations, a subproject headed by Ellen Bass, Leigh Baumgart, and other ERC-CASA team members sought to better understand how emergency managers make sense of, manage, and apply radar data in the short-term evaluation of severe weather probabilities and warning decisions, the goal of which was to align and integrate ERC-CASA's DCAS system more closely with the realities of potential end-users. The experimental scenarios presented a series of mock exercises in which a variety of stakeholders—in the case of CASA, primarily emergency managers and forecasters—made decisions based on pre-existing radar data (e.g., wind velocity, reflectivity) so that their behavior, needs, and challenges could be observed and analyzed for patterns throughout the exercise.

The end-user simulations revealed several important insights relevant to the interdisciplinary aims of CASA. First, not all users held the same training and expertise to make full use of the DCAS system and data. Emergency managers generally fell into one of two groups defined chiefly by expertise and familiarity with radar data: *conventional* users and what were termed "*super users*" (Bass et al. 2009). Conventional users generally held limited training, and therefore limited familiarity with the nuances and features of radar data, whereas super users, while not weather experts strictly speaking, shared a broader familiarity with weather and radar concepts above and beyond what was observed among conventional users. Noticeable differences in expertise came into play when emergency managers were asked to interpret velocity data, which is of a more complex nature in comparison to reflectivity data. Emergency managers appeared to struggle to interpret the meaning of velocity data—conventional users especially so—as velocity data is far less intuitive relative to reflectivity data and requires the interpretation of multiple variables and variables related to immediate weather patterns. These results were particularly relevant for CASA and the design of the DCAS system: if even earlier radar generations could produce a level of data complexity too demanding for a considerable number of likely DCAS stakeholders, it follows that CASA radar, by design more intricate, may have created natural difficulties for the novice or untrained emergency manager, of which there were a considerable number in Oklahoma. The simulations also revealed that emergency managers confronted problems managing multiple sources of information and making use of such information in a timely manner (in some cases, emergency managers overlooked dangerous, emerging storms due to the complexity of data). Bass et al. (2009) summarize the problem accordingly:

> When presented with multiple radar products at varying temporal and spatial specificity, such as with WSR-88D and CASA data, EMs must know how to effectively integrate the information. For example, more than half of the EMs commented that they use WSR-88D 12 reflectivity to keep the big picture of approaching storms to the west and south. However, when presented with DOW data, one EM was so busy using the DOW data to look at the storm passing over the town, other storms moving in were ignored. Emergency management training through OK-FIRST is currently centered on five minute update rates of WSR-88D reflectivity and velocity data. With CASA update times potentially five to ten times greater, EMs may benefit from additional training related to time management. (11–12)

How does this connect to the development of CASA's technology? Most importantly, this establishes potential problem areas when the technology is made widely available, areas for which training and education is most needed within the technical end-user group. In this case, if the DCAS system is to provide complex imagery, such as velocity data, for such data to be of any use emergency managers will need additional training in this regard. More importantly, the faster updates (note that WSR 88-D only updated every 5 min) from 5 min to quicker of weather information may not have been useful because even at 5 min updates storms were missed. Thus, the interdisciplinary approach called for more training in this regard.

Policy Development

As noted in Philips et al. (2007), among the many proposed innovations, another novel feature of DCAS was that it allowed for the real-time manipulation of the distributed radar network when needed (hence, the "adaptive" and "collaborative" terms in the acronym—the features were informally referred to as "joy sticking" the system). This means that the system, if necessary, could if so commanded reorient sensing and computing resources to areas in the environment where most needed. For instance, a storm of growing size and increasing circulation might merit greater attention and system users could readjust radars to focus on the event accordingly. In a normal state the radars would simply spin at a standard rate and with standard power, an area in which a storm was developing could be granted more power and resources so that a more detailed picture of the storm event might be painted.

Thus, the system was designed in such a manner as to respond to the needs of technical stakeholders (once again, forecasters and emergency mangers). However, as seen with all applied science, the approach held several untested assumptions about system use. Notably: that among users the common factor deciding how system resources would be focused would come down to the size and intensity of the storm system. In other words, if two storms systems appeared in the DCAS network, the larger and more violent of the two would be "targeted" for observation through the reallocation of system resources. At face value, the assumption would not appear unreasonable, but from a sociological and psychological standpoint, the matter is far from straightforward. In keeping with the spirit of interdisciplinary, we should ask: *should "joy sticking" be simply a sole function of meteorological variables* or are there other important factors relevant to who should control the system and what variables should govern what storms are more heavily scrutinized? The answer was a complicated one. A third thrust in the end-user interdisciplinary integration effort concerned the development of an overall policy by which the DCAS system could be managed and controlled. This involved interviews and focus groups with emergency managers and forecasters.

Interviewed emergency managers and forecasters suggested that while storm size and intensity would be critically important in directing system scanning, there were several caveats to this policy, the most prominent of which related to known "gaps" in coverage in smaller media markets (Rodriguez et al. 2010). As noted in Chap. 4, media institutions tend to focus on larger markets due to the financial advantages in doing so. Thus, areas with larger populations—notably urban and larger suburban areas—tend to receive in relative terms greater and more sustained coverage during periods of severe weather. For instance, Norman, OK, due to its larger market size, would receive more coverage from local media in comparison to more rural, sparsely-populated areas of the state. As a consequence, a moderately-large storm forming over rural communities may not receive as much coverage as perhaps a less-intense storm over Norman or Oklahoma City when market considerations factor into the decision of regional coverage.[2] Thus, there may be significant gaps in media attention to storms developing near smaller communities when a storm system is large enough to potentially threaten a wide range of smaller and larger markets. Some emergency managers reasoned that the ability to joystick the DCAS system could fill some of these gaps in coverage for smaller communities since large communities would already command a significant share of attention from public institutions in the face of severe weather. In other words, the ability to joystick the system might even the playing field as far as severe weather is concerned.

This was not the only factor that shaped the anticipated dynamics of the DCAS system. Rodriguez et al. (2010: 35) suggesting a host of considerations shaped perceptions of how the system should be joysticked:

[2]This is perhaps an aside, but this illustrates a key tenet of the Sociology of Disaster: that disasters are not necessarily natural. Rather, what makes a disaster is how society *responds* to an environmental threat.

(a) nature of the hazard event, (b) potential impact and outcomes of the hazard event, (c) lead time, d) false alarm rates, (e) population issues, (f) infrastructure, and (g) availability of other resources.

As far as the DCAS joy sticking was concerned, interviews suggested that decisions regarding how, when, and where to direct system resources was likely to be a complex and multidimensional affair, made on the basis of a variety of considerations and evolving observations, some relatively static (e.g. the presence of infrastructure, high population density) and others more dynamic (e.g., nature of the hazard events, anticipated impacts). No single factor would determine system use, it seemed. On the contrary, factors considered to be relevant were to be continually weighed against one another in the decision to allocate and reallocate DCAS scanning foci. For instance, in the context of a hypothetical DCAS joy sticking scenario, a forecaster might initially focus radar scans on a moderately large tornado evolving near a sparsely populated region, but, according to findings, later turn system attention to a weaker storm system threatening a high-population area. Judgments with respect to the use of the system were relative to a variety of competing factors and were always subject to change should the evolving circumstances characteristic of severe weather outbreaks call for a revision to scanning policies.

Another key matter involved the question of who would control the DCAS "joystick." Put differently, who would best be in a position to hold final jurisdiction over the dynamic controls of the DCAS system? Were forecasters, with their training, expertise, and education most suited to helm the network? Or were emergency managers, who had more localized knowledge of storms and communities, as well and a more immediate ability to assess local severe weather scenarios, in a better position to make use of the adaptable elements of the system? If given to forecasters, the system might default to weather models, losing a sense of local context. Emergency mangers, on the other hand, witness dynamic, local changes in weather patterns, and therefore may be in a position to adjust the system accordingly. Yet it was ultimately determined that emergency managers may not have the training to properly assess these changes, and so forecasters were favored to have direct control in the final analysis.

Public Response Research

While the previous sections focused mainly on research intended to integrate technical end-users, another key element of CASA was a focus on the integration of what were previously branded "beneficiary end-users"—those for whom the system is some benefit, but who at the same time play virtually no role in the direct scientific or technical direction of a project. In the case of hydrometeorological threats, of which was central concern to ERC-CASA, the principal beneficiaries—those at the end of the stakeholder chain, so to speak—are, by and large, people and communities. Acknowledging and integrating beneficiary end-users is absolutely critical for an interdisciplinary project—ERC-CASA or otherwise—because the ability of systems to benefit communities often directly relates, once again, to the "global question" or *useful function* of any interdisciplinary research effort. In the case of CASA, recall the global question was the following: how do we mitigate severe weather fatalities, injuries, and economic losses through advances in remote sensing technologies? Even if technical end-users—in CASA's case, emergency managers and forecasters—were to be successfully integrated into system design, such that the DCAS system were to improve decision making at the institutional level, this would mean very little at the end of the day if the new radar system ultimately held little value in the form of improved safety for the public. The matter is uncomplicated: *simply because DCAS was superior to radar technologies of the day did not mean it was also superior in terms of saving lives.* Remote sensing, forecast data, and prediction algorithms, however advanced, are all simply too far removed from the basic realities that govern public response to be of much *direct* use to the community. Although there is little research on the topic, it is perhaps not unreasonable to assume that the general public knows little about remote sensing, radar technology, or radar data—at least not well enough to use them directly as a warning tool.

Thus, the final end-user research thrust involved development of public response fieldwork and surveys through which a "baseline" of public response behaviors and attitudes were defined in relation to the warning response model outlined in earlier chapters (recall the model introduced in Chap. 3). While findings presented in the next section have been discussed elsewhere at great length and in far greater detail,[3] here these data are summarized with an eye towards illuminating the interdisciplinary elements of CASA and how public response research informs system design.

Overview of CASA's Public Response Fieldwork

Data collection via qualitative fieldwork was conducted from February 2006 until April 2006, within which time three separate tornadic events in Louisiana, Missouri, and Tennessee formed the basis for an on-site, quick-response study of public warning response. Each fieldwork site was highly distinct geographically and demographically. For instance, fieldwork in Louisiana took place in a diverse suburban and urban area around New Orleans, whereas fieldwork in Tennessee progressed within rural areas defined by greater demographic homogeneity. More specifically, these areas included New Orleans and surrounding Jefferson parish in Louisiana; cities and towns in the regions surrounding Springfield, Missouri, as well as rural communities in the immediate locality of Dyersburg, Tennessee. The approach was by no means accidental. When studying beneficiary end-users, such

[3] For a broader theoretical treatment, as well as methodological description, see Donner et al. (2012) and Donner (2007). The sections that follow to the end of the Chapter are adapted, revised, and applied to the interdisciplinary discussion from dissertation work in Donner (2007) with the new purpose of illustrating interdisciplinary concepts.

diversity of environment, population composition, and geography is deeply important since the goal is to make technical innovation as useful and applicable to the broadest range of stakeholders as possible. Altogether, 59 interviews took place in the areas of New Orleans, Springfield, MO, and Dyersburg, TN, which returned a highly varied sample of people and places.

New Orleans: Disaster in the Context of Recovery

New Orleans, as an urban area, and one having been recently impacted Hurricane Katrina, presented the most unique context in which to study warning response. For one, major cities generally face greater levels of vulnerability for two reasons, one of which is demographic: large urban areas have been the traditional home for society's most vulnerable groups—racial and ethnic minorities, the poorly educated, the impoverished, and the isolated elderly, all of whom face difficulties in preparing for and responding to disaster (see Chap. 3 once again for a broader overview). The second reason cities face significant risks is related to the nature of their structure and development. All else equal, cities are difficult places from which to evacuate due to the potential for high rates of traffic congestion and the prevalence of asphalt and concrete (two highly impermeable surfaces), which, together, makes all the more likely conditions for catastrophic flooding. Indeed, when Hurricane Katrina struck New Orleans in late August, the dual influence of both demography and the structure of urban environments would be felt not only in New Orleans, but across the Gulf Coast. In the wake of Katrina, CASA researchers set out to study a series of unexpected tornadoes for which the local community was largely unprepared and to which was therefore highly vulnerable.

New Orleans has historically been home to a large African-American community, who, according to the National Census, made up at the time nearly 70% of the city's total population prior to Hurricane Katrina [see Table 7.1 from Donner (2007)]. At the same time, the city is well above the national average for individual and family poverty rates. According to a report issued by The Urban Institute using 2000 U.S. Census Data, at the time Katrina struck, New Orleans suffered a staggering poverty rate nearly 16% higher than the national average, while at the same time housing one of the poorest black populations in the nation (Urban Institute 2005). To make matters worse, New Orleans is located at one of the lowest points in Louisiana and one of the lowest points in the United States, which created a natural "bowl" within which water collects.

When the tornadoes struck in February, 2006, a combination of demographics, geography, and ongoing recovery efforts from Hurricane Katrina made for a highly vulnerable population in New Orleans and surrounding areas. Making matters worse, New Orleans rarely experienced severe weather beyond Hurricanes, resulting in communities largely inexperienced with the unique risks posed by tornadic storms. Such lack of familiarity, fused with the characteristic vulnerabilities of the area and the recentness with which Katrina struck, would be tested in early 2006: on February 14 of that year, several strong tornados hit the city, as well as its adjoining

Table 7.1 Demographics of sampled communities at the time of study

	ST	Population total	% African-American	% Hispanic	Med age	% HS Grad	% Disable	% Fam Pov	% Ind Pov
Kenner	LA	72,874[a]	25[a]	13.6	35.6[a]	81.1[a]	14.9[a]	11	13.7[a]
New Orleans	LA	437,186[a]	67.5[a]	3.1[a]	35.2[a]	82.3[a]	15.6[a]	21.8[a]	24.5[a]
Hahnville	LA	2792[a]	50.9[a]	1.1[a]	34.2[a]	75.5[a]	18.6[a]	19.8[a]	21.6[a]
Nixa	MO	12,124	0.5	1.3	31.9	88.5	16.5	8.1	9.7
Battlefield	MO	2385	0.2	1.2	32	90.1	16.5	1.5	2.4
Republic	MO	8438	0.2	1	33.3	82.2	17.9	5.8	6.7
Marionville	TN	2113	0.1	0.6	38.5	75.6	26	19.5	22
Fordland	TN	684	0	1.6	34.3	79.4	26.6	4.8	11.4
Verona	MO	714	1.5	31	30.8	60.4	22.9	19.4	26.4

[a]Estimate prior to Katrina. Possibly not accurate estimate of population at the time of study, especially in Kenner and New Orleans. Source: U.S. 2000 Census. From Donner (2007)

Fig. 7.5 Manufacturing plant destroyed by tornadoes in New Orleans

communities, causing severe destruction to an already devastated region. Most of the destruction—resulting from two or three F1 tornadoes—occurred in Kenner and Metairie, the former a western suburb of New Orleans, where severe damage was done to Louis Armstrong International Airport. Metairie, more suburban by most standards, housed a large manufacturing plant as well as several storage facilities, all of which were damaged by the tornadoes (see Fig. 7.5). Most damaging to the region, however, were the tornadoes that struck the airport. As a result, "Electricity was knocked out at Louis Armstrong International Airport, grounding passenger

Fig. 7.6 FEMA trailers were a common sight in many gulf coast communities at the time the tornados struck

flights and leaving travelers to wait in a dimly lit terminal powered by generators. The storm also ripped off part of a concourse roof, slammed one jetway into another, and flipped motorized runway luggage carts" (Associated Press 2006). The American Red Cross reported that tornadoes damaged roughly 300 houses (Red Cross 2007). Other touchdowns occurred in Lakeview, a then nearly uninhabited area hard hit by Katrina, as well as a small suburb south of the river. No deaths or injuries were officially reported, but a number of those with whom interviewers spoke revealed minor injuries such as cuts from glass (Fig. 7.6).

Missouri

The second fieldwork trip took place in Missouri at locations where from March 9 through March 12th, 2006, a major tornado outbreak struck as part of a broader outbreak in the central U.S. As a result of the storms, authorities accounted for more than ten deaths and approximately $1billion in damage to residential housing and businesses across the southwestern United States. Fieldwork took place in Springfield and the surrounding communities of Nixa, Marionville, Battlefield, and Republic, Verona, and Fordland.

It should be noted that in comparison to New Orleans, tornado warnings and watches were far more common in southwestern Missouri, resulting in a far stronger awareness and appreciation of tornadic threats on the part of the public, but, as one interviewee put it, as well a "numbness" among many residents to the warnings and

Fig. 7.7 Destroyed housing in Nixa, MO

watches regularly issued by emergency management officials and NWS. During the outbreak, a total of 84 tornadoes spawned from a large supercell causing extensive damage across the region. In Nixa, a small town near Springfield, three injuries were reported along with 40 damaged homes. Nixa at the time was one of the region's more affluent housing developments located northwest of Jefferson City and interviewers encountered several damaged homes (see Fig. 7.7) and bare foundations (see Fig. 7.8). Similar search and rescue nomenclature to that observed during Katrina was spray painted on some of the houses, indicating whether the house was searched, residents found dead and so on. In Nixa, residents generally recalled the sirens being set off 15–30 min prior to the tornado's onset.

Other municipalities experienced substantial damage, as well. Near Battlefield, a smaller community in the region, there were four injuries along with downed power lines, downed trees, debris damage, and homes destroyed near the major highway. Republic, another small community in the Springfield area, experienced the lightest damage of the five communities researchers visited.

Marionville, a small city located in the southwestern region of the state, by far experienced the most extensive damage: there were three injuries, one fatality, and numerous houses destroyed. Damage patterns were similar to that observed in Nixa: collapsed roofs, toppled mortar and brickwork, and bare foundations. Unable to withstand tornadic winds, sheds were also commonly destroyed as were many outside structures. Marionville is an interesting case because the recent tornadoes took similar paths to those that hit the community in 2003 (a mere 3 years earlier), several interviewees recalled. The 2003 tornado struck similar areas as the 2006 one had, resulting, according to one interviewee, in greater degrees of preparation and awareness among local residents. In Verona, a community of approximately 1 h

Fig. 7.8 Foundation of home, Nixa, MO

distance from Springfield, there was one fatality and one injury with a one-mile track extending northeastward through the community. Of demographic concern, here a sizable minority of residents were Hispanic (see Table 7.1), many of whom are employed at a local poultry plant outside of town. The white residents with whom interviewers spoke held positive views towards the Hispanic presence, saying that many had actually refurbished downtrodden homes and in doing so raised local property values.

Geographically, the areas researchers visited lie within or close to the Ozark Mountains (in reality a misnomer as a majority of the Ozark region consists of highlands or plateaus), which is composed of the Springfield Plateau, the Salem Plateau, and the Boston Mountains. As will be seen, the importance of Southwestern Missouri's geography, particularly with respect to the role of disaster mythology in warnings, would emerge within the interviews.

Tennessee

Finally, in April, 2006, researchers visited western Tennessee following a major tornado outbreak that claimed several lives (see Fig. 7.9). Interviews were first conducted in Millsfield, TN, a small town approximately 15 min east of Dyersburg, one of the larger cities in northeast Tennessee. Upon arrival to the small rural community, researchers encountered sporadic, but unignorable, damage to housing and trees. A foundation, where a church once stood, sat on one side of the street. It appeared as if several other buildings fared little better during the storm, suffering

Fig. 7.9 Abandoned housing near Dyersburg, TN

extensive damage. Adjacent to a destroyed firehouse, interviewers noticed a deep swath cut into the hillside, ostensibly by a tornado. Millsfield, at the time, had no siren system of its own, so sirens on fire department vehicles are generally deployed. Unfortunately, the firehouse was destroyed during the event, for all practical purposes eliminating what was one of the most important sources of warning to the community.

According to reports, the tornado touched down in Millsfield around 7:30pm. It would appear, therefore, based on the time at which the tornado struck and the discussions with interviewees, that many residents were at home. In Newbern, a community approximately 10 miles from Millsfield, there were several mobile homes fortified with a layer of brick, as well as one- and two-story houses. The closest siren network is in Dyersburg, located approximately 10 miles to the southwest, and, according to one interviewee, is rather difficult to hear.

Another important research site was Caruthersville, TN, where there was extensive destruction of housing and other buildings. Large sections of a local high school had collapsed and many windows were shattered. Immediately noticeable upon arrival were demolished homes with debris covering the street, most of which consisted of the countless trees that had fallen during the storm (see Fig. 7.8). Caruthersville had clearly been hard hit by the tornadoes. To illustrate this point one need only refer to the fact that FEMA established temporary housing on or near the foundations of destroyed housing. In neither New Orleans nor Missouri had FEMA's presence been so evident. As far as the warning system is concerned, the fire chief, along with the mayor and police chief, hold the authority to issue local warnings.

As far as infrastructure is concerned, Caruthersville is unique in comparison to other research sites in Tennessee and elsewhere. Most notably, Caruthersville is home to a floating casino docked on the Mississippi river. Interviewees on the boat and the bartender said that during the storm gamblers would not seek shelter: settling gambling debts and ordering more beer was prioritized over protective action. The bartender, who was ushering people downstairs, later ascended to the outside deck with a co-worker in order to visually identify the tornado. The Mississippi river, too, would play a significant role in shaping residents' perceptions of the warnings they received that day in April.

Synopsis of Results

While a more comprehensive overview of the results and methodology from the public surveys may be found elsewhere (see especially Donner 2007; Donner et al. 2012), here are presented a brief overview of findings that illustrate the results from the fieldwork. Here results are discussed with an attempt to provide a model of integration in to the ERC-CASA DCAS system in an effort to broadly illustrate the interdisciplinary approach.

Reception

Notably, reception was found to occur primarily through two channels—informal and formal networks. While it was not uncommon to receive official warnings, many respondents reported receiving warning information through neighbors or other informal channels of communication. As one interviewee noted: "Probably...I'm not even sure we received the watch as much as neighbors calling, and I think they just received it and we were getting prepared for bed and all cuz it was I believe it was at 10:30 at night, so we were I think that's when it hit, so we were kind of at that point, so" [QR-MO-6]. Here the fundamental lesson in the context of ERC-CASA was that a considerable number of warnings may never reach members of the community through formal sources such as the NWS or local emergency management. For instance, according to interviews, undocumented Latinos in one community rarely received warnings directly, but rather through local religious leaders. What are the policy implications of this observation from an interdisciplinary perspective? Since minority groups are less likely to receive warnings, a cost-effective way to reach these groups is through the informal institutions and groups often turned to when information is necessary. For Hispanics, trusted institutions with Spanish speaking leadership and members such as churches, local non-profit organizations, and volunteer groups could conceivably serve in this role.

Further barriers to reception were found to exist in the sample. Hearing disabilities, poor or dated siren infrastructure (e.g., "And they set [the sirens] off every Saturday at 12 o'clock. But you gotta turn the T.V. down if you want to hear it" one interview lamented [QR-TN-11]), isolation from personal and group networks within the community (though which warnings are communicated), and lack of

English commonly accounted for missed watches and warnings. Nocturnal tornadoes were found to present a unique dilemma: "It was like the middle of the night. So, *I didn't wanna wake anybody up*. . .had I heard the train whistle, well maybe I would have called, but I didn't hear anything" [QR-NO-18]. Again, going back to earlier discussions, there was the question about whether the anticipated risk of tornadoes exceeded the social costs of waking family members up in the late hours. Norms were accordingly found to govern the transmission of warning information between members of the community.

Understanding

After a tornado warning is received, as predicted by the warning response model, respondents attempted to make sense of message content—to scrutinize, precisely, what the message meant and what the senders intended to communicate. What were the location and intensity of storms? What am I being told to do and how am I being told to respond? It is instructive to note that while respondents held a general comprehension of warning language—at the time of the case studies severe weather "warnings" and "watches" remained the dominant vernacular through which the public received tornado alerts—oftentimes the nuanced meanings between warnings and watches were lost in translation. This confusion is important as either term conveys meaning of a radically different character. To wit, watches typically communicate the *potential* for severe weather, whereas warnings make clear tornadoes have formed, touched down somewhere reasonably close, and are of a direct threat to life and property. Some interviewees could distinguish between the two, but others struggled to define either in distinct, pragmatic, and ultimately actionable terms, whereas others simply confused the two. A common error in comprehension reflected a tendency to take warnings as watches (and vice versa) and definitions themselves were often vague and uneven: "*a warning, as I understood it, was such that it was to the point that a tornado could develop at any time*" [QR-MO-2]. Of course, this is not the correct definition of a warning, as warnings *meant* a tornado was already on the ground. Adding to the confusion, in communities in which sirens were used to warn multiple events, interviewees appeared to have trouble arriving at a consensus as to exactly what the sirens meant when sounded. These problems were more pronounced in Louisiana, where tornados were rarer events. For instance, in one community, interviewees, even in the same household, debated over whether local sirens were meant to signal a chemical leak from a local plant or a tornado. As it turns out, the sirens were intended to signal both.

Here the dynamics of the model were found to emerge. As noted earlier (recall the end of Chap. 3), the stages of the model are fundamentally linked, i.e., what takes place in earlier stage does not occur in isolation, but rather has impacts on later stages in the model. In the case of the transition from understanding to belief, this principal was clearly apparent: "*we watched and listened...because we've never had anything like this and so we had had never became all that concerned*" [MO-NO-2]." In this case, because the interviewee, due to lack of experience, had difficulty making sense of tornado reports and warnings at the time of the storms, the result was an inclination to disbelieve the warnings. Tornadoes can be, it seems, so incomprehensible that that they were *unbelievable.* This adds evidence to the idea of a dynamic, rather than static, model of warning response, one in which the stages cannot be regarded as simply separate from one another.

Belief

As was expected given what is known about risk communication, it was perhaps unsurprising to discover that many interviewees greeted tornado warnings with a fair degree of distrust. In many cases, especially in the Missouri and Tennessee case study sites, but perhaps far less so in New Orleans, the general public regarded tornado warnings merely as a part "everyday life" rather than something extraordinary. It is interesting to note that several interviewees seemed to treat warnings with a sense of annoyance and irritation rather than urgency. Warnings *themselves* accordingly seem to do little to disrupt the prevailing sense of normality through which the world operates. Normalization tended to eclipse any sense of urgency created by warning messages, making them largely ineffective in most cases. This tendency towards normality leaves many unmoved as to the possibility of danger for two reasons.

First, as a result of overexposure to tornado coverage, particularly during tornado season, respondents reported that warnings in many cases evoked passing concern at most. Warnings were so common as to be regarded as unremarkable and were therefore met with a noticeable degree of apathy among interviewees. Constant and unremitting weather reports throughout tornado season may lead to a certain psychological fatigue that in turn brings about a sense of indifference towards the risk of severe weather. Such normality may also grow over time and be further reinforced as the public continually witnesses community institutions simply carry on operations, as if everything was normal, as warnings are issued: "[During the warning, the employees] just kind of kept doing our work, and we still had people coming in and ordering food, and nobody really did anything" [QR-TN-12]. Social cues indicating the presence of heightened risks were noticeably absent—indeed, the continued operations of local business led interviewees to believe everything was normal. Indeed, as a constant feature of everyday life, one may simply not have the time to respond to each and every single warning, alert, and weather report or life itself would be impractical.

While still a matter of debate in the literature, false alarms nevertheless factored heavily into discussions about the credibility (or lack thereof) of tornado warnings, but in a manner that was neither simple nor straightforward. Some respondents create psychological benchmarks—again, heuristics—as a method by which to evaluate the truth of tornado warnings: "Um you might hear [warnings] um you know 4 or 5 times a year, something like that. You're under a lot of watches, obviously, but um you might have more than that but typically you kind of know by which area of Springfield that [the tornados] are heading. If it's going north, you don't even think about it. You just kind of turn up the TV and ignore the sirens" [QR-MO-6]. The need for heuristics, once again, may result from the oversaturation of weather information. This led some to some to rely on what might be called "metaphorical cues" as a heuristic to determine whether to take warnings seriously. For instance, a common metaphorical cue among interviewees was that tornado threats were present was when a "train whistle" was heard [QR-NO-18]. Similarly, in Tennessee: "The siren goes off so many times, we had sirens go off and a tornado no where near us, it was in Dyersburg but you know they spotted one in Dyersburg so the sirens go off here to warn us to take cover. Well, you know, I'm not gonna go lock myself and my husband up in a closet for an hour wondering if this thing's passed [QR-TN-11]." Warnings, by themselves, are necessary but insufficient to erode the sense of normality that governs attitudes towards tornado warnings.

Confirmation

Noted earlier, absent stark environmental or social cues, warning recipients often elect to confirm severe weather watches, warnings, and reports. At this stage of the model, warnings are suspected to be true, but enough doubt remains to make recipients unsure whether taking action would be worthwhile (once again, warnings are simply a part of everyday life for many in tornado-prone regions). Figuratively speaking, confirmation builds a psychological bridge between the earlier stages of the model—that is, reception, belief, and understanding—and later stages starting with personalization through protective action. In the ERC-CASA fieldwork study, it was the rare interviewee who did not confirm warning information. Yet, while the tendency to confirm was found to be common, the means by which confirmation takes place appeared to be less so. One may elect to turn on the television or contact friends. Others may go as far as to look outside to visually confirm the presence of severe weather. Through any method of confirmation, however, immediate preparation and sheltering experiences delay, an important consideration given that "time is of the essence" as far as severe weather is concerned, especially tornados. Worse still, in some cases the act of corroborating storm reports actually places recipients at greater risk: "Cause you could see 'em" one interviewee reported, "about right over there. Yeah you could see about 300 yards, coming up this way," implying that the interviewee visually confirmed tornadoes after receiving a warning, which is the opposite behavior of that intended by the warning message.

What is more, the initial clarity and precision of initial reports, it would seem, had little bearing on the impulse to confirm. In one case, an interviewee even recalled in unambiguous terms the warning's instructions: "they just said on the TV that if anybody had a place to get like in a basement, a hallway, or a place with no windows, do so immediately", yet the same interviewee still indicated seeking confirmation of the storms. This is in spite of the fact that the language of the warning, she appears to remember, clearly called for action, implying dire, threatening circumstances. Indeed, the interviewee *believed* tornadoes were forming, and she held few doubts about the presence of severe weather, but in the end succumbed, despite her initial credulity, to the need to confirm that the tornadoes presented an immediate and unequivocal threat to her and her family—that the tornadoes were not merely a general risk, but a *personal* risk (once again, illustrating that confirmation serves as a bridge between general belief and risk personalization). It took repeated warnings witnessed on television before the interviewee eventually decided to seek shelter. The need to personally see or experience the tornado or the effects of the tornado also factored heavily into the confirmation process. Note in the following exchange a description of the detailed visual inspection of surroundings performed by the interviewee to evaluate whether or not an immediate danger existed:

> *Interviewee*: The sirens and then people talking. And I do know that across the street that, well actually I had seen the tornado and the front was coming in on the other side of Caruthersville, so I kind of knew that I thought it was going away from us.
>
> *Interviewer*: Oh so you were coming from Caruthersville and you saw a tornado down there.
>
> *Interviewee*: I saw a tornado in Caruthersville and usually the weather goes north of us and it was sunny here, right before the tornado hit. [QR-TN-5].

Personalization

Of the all the stages, personalization was revealed as arguably the most complex and critical insofar as it acts as a sort of threshold between previous psychological attitudes towards the warning (e.g. understanding, belief) and behavior (e.g. "protective actions"—sheltering, seeking refuge, etc.). Simply put: *personalization* occurs when the process of confirmation converts the belief that there is a tornado risk somewhere "out there" into a belief that one is in *immediate* danger. The process of personalization is far from straightforward, it was discovered. For instance, when presented with two different severe weather events, some interviewees held a tendency to personalize the threat perceived as most immediate regardless of the inherent danger presented by either. To illustrate this point, consider the following statement made by one interviewee: "Right, well that's what I thought. Matter of fact when I heard that there was hail, my mind said, 'oh hail instead of tornado.' Which makes it okay to run out and move your car cause I'll brave the hail, not an F-3 tornado 500 feet away." [QR-MO-8]. Though the respondent clearly understands tornadoes to be more dangerous (*I'll brave hail, not an F-3 tornado 500 feet away*) in the above quote, the interviewee acknowledged the threat of hail as more *relevant* to his personal circumstances and as a result largely ignored the threat of tornadoes (despite the fact that tornadoes were mentioned in the message).

Local culture furthermore shaped whether one personalized a tornado warning or not. In this regard, interviewees referenced beliefs about the role of local geography in offering protection from tornadoes. For instance, if it was believed that mountains shielded the community from tornadoes, one was less likely to personalize the risk communicated in the warning. Similarly, rivers were thought to afford protection and, regarding urban settings, one respondent remarked: "I never heard of that [it could come through a city] you know being in urban areas" [QR-NO-13].

Interviewees often used what we term here "geographic benchmarks" as a means to evaluate personal risk. Observed impacts to neighboring communities often served as a psychological yardstick by which one's own personal risk was evaluated: "When we saw the storm was, was going through this other community of Billings and I knew that the storm. . .would just. . . if it missed us by any, it wouldn't miss us by much. We decided just to go ahead and go to the shelter, where it's partially underground and we kind of just kind of felt more comfortable under there" [QR-MO-5]. Once again, it was through these processes that risk moved in the minds of respondents from something abstract and intangible to an immediate, real threat to life.

Determining Whether Protective Action is Necessary and Feasible

As noted earlier, between "Personalization" and "Protective Action," there exists one final psychological stage through which warning recipients are observed to pass before action may take place: the evaluation of whether action is actually necessary to reduce a personalized risk or whether no action is needed. Once again, this may sound strange, given that one might naturally believe that a personalized risk would be immediately responded to, but, as Mileti's model has suggested, and as was observed in the interviews, some may feel powerless to respond or believe—whether correctly or incorrectly—that present circumstances afford sufficient defense against risks however immediately perceived. Many factors shaped evaluations of the necessity of taking action in response to risk personalization.

While far from a universal topic of interview conversation, fatalism played a somewhat considerable role. At the end of the day, some interviewees simply felt powerless to protect themselves, which, in turn, led to inaction: "I remember, all I remember him saying is tune in and watch. If you're in—if you're one of the residents in the FEMA trailers, you know it's—you must leave and we were all laughing because where are we gonna go. I mean the whole reason we're all living in these trailers is because we don't have any houses. So it was like a joke, you know. So where are we gonna go? You can't even get a hotel in New Orleans, you know" [QR-NO-13].

Past experience and knowledge also factored into decision about whether response was necessary. Again pointing to the earlier-discussed *dynamic* elements of the model, while the literature suggests a familiarity with disasters improves understanding and lessens skepticism, familiarity plays a far less straightforward role with respect to whether one regards the act of response as *necessary*. In the case of CASA fieldwork, there was a notable tendency among interviewees, especially those in New Orleans, to apply common hurricane response strategies to tornadoes, which is altogether inadvisable. Applying hurricane response strategies (e.g., evacuating, sheltering in attack to avoid flooding) is the exact opposite of what one should do to protect oneself from tornadoes (e.g., basement sheltering). Indeed, several cases were found in which interviewees misapplied response strategies. In one instance, a New Orleans resident took shelter in his home's attic in response to tornado warnings in replication of what was done several weeks earlier during

Katrina. It may be since this strategy worked for Katrina, the interviewee, in turn, thought it would work for tornadoes. Without the presence of a strong "tornado culture" in New Orleans, the interviewee simply and understandably defaulted to the response strategies drawn from the dominant "hurricane culture" with which he was most familiar. Indeed, at least three interviewees speculated that they may have evacuated had they known about the tornado or would evacuate in the event of future tornadoes (again, evacuation is appropriate for Hurricanes, but not so for tornadoes). One reported that "[following] *Hurricane Katrina, [we] lost everything. If we'd had a warning, trust me, we would have got out of here. Took what we could take and we woulda roll*" [QR-NO-1].

Protective Action

If determined to be needed and feasible, action is then taken to protective oneself and, more often than not, those within one's primary group. It should be noted that the actions one takes in response to warnings may either reduce the risk one faces or place one at greater risk—it should be emphasized that the model is a descriptive and not a prescriptive one. For instance, one may regard driving away from a tornado as both needed and feasible, but doing so likely places one in greater danger, thereby resulting in an outcome opposite as projected. There is one important question, however, that remains of the warning process: how do people decide *what* course of action to take and what influences this decision? *Here is discussed how people act when responding to a warning.* While there were many different patterns into which protective action fell, two factors consistently oversaw the process of protective action: *family/network structure* and the *availability of resources*.

1. *Primary groups (e.g., family, friends) will attempt congregate during warnings.* Regarding protective action, the literature tells us that there are few lone evacuees (Aguirre et al. 1998), an observation that is largely uncontested by the CASA fieldwork data. By way of illustration, one interviewee reports: "She said she thought that.... she has a big closet in her bedroom (mm hmm) that would be a safe place for her to go. So she wanted to know if she could do that, and I told her 'yeah' if that's where she wanted to go, that would be fine, but she would have to help me into the bathroom because I couldn't walk far enough to get in her bedroom, into the closet. But then she chose to stay in the bathroom with me 'cause she didn't want to be alone" [QR-MO-4]. What must be understood is that even if there is no *instrumental* need to assemble during warning periods, and barring any obvious and immediate risks to doing so, congregation is common as a means to provide the comfort necessary to psychologically endure the trauma of a major weather event. Indeed, there were many reasons why interviewees contact—or were contacted by—members of primary or close secondary groups: to make sure others were aware of warnings, pool resources, check on the safety of loved ones, and offer suggestions about what to do if the tornado were to strike, among others. It should be noted that while members of secondary groups—such as neighbors and co-workers— were sought out in some cases, primary group

members played a more consistent and sustained role throughout the warning process—from evaluating warning information, negotiating the feasibility of protective action, and gathering in an effort to better survive the storms. Interaction among secondary groups sometimes consisted of offers to use basements and other shelter or to briefly exchange information regarding storm reports. This was particularly evident in Nixa, a small housing development in Missouri, where interviewees with shelters offered refuge to neighbors during the tornadoes. One interviewee told researchers that he informed his neighbors that the shelter was opening [QR-MO-5]. In other cases, if one did have the resources to protect oneself, these resources were offered to others in the community: "*And we also told the neighbors on the north side of us who do not have a basement that if they felt imminent danger they were more than welcome to come over and join us in our basement*" [MO-NO-2]. This is not surprising since altruistic behavior during disaster has been near axiomatic within the field of the Sociology of Disaster since inception.

Taken together, these observations suggest that members of a primary group separated at the time of a warning will attempt to reunite if possible outside of only the most highly dangerous circumstances. With that said, however, the definition of a "highly dangerous situation" may vary from person to person, and the boundary between acceptable and unacceptable risk often blurs during periods of warning, particularly when loved ones are involved. There were, as such, situations in which interviewees claimed to have united with family members despite having already having heard sirens. The importance of primary ties in one's interpretation of and reaction to risk cannot therefore be overstated, a fact which gives further support to the notion that warning response is shaped by social norms rather than an instinctual response to risk, a point portrayed in a case involving an elderly female interviewee in Newbern, TN. She noted during discussion that her grandson rushed to her house after hearing warnings and also being fully aware that tornadoes were rapidly approaching town. Social bonds, particularly among members of primary groups, are unlikely to be abandoned or ignored in the face of mere risk.

2. *Infrastructure Available.* Finally, one obvious barrier facing storm victims is the lack of safe shelter—most commonly in the form of a basement or a private tornado shelter, or in other cases a public shelter to which schools and public buildings are often converted in times of severe weather. Many simply could not seek refuge from storms due to the absence of accessible shelters. In some cases, public shelters were sought out, but many lacking a safe space to hide from the storms simply elected to shelter in closets and bathroom in the absence of any alternative.

Conclusion: Integration and Connection to System Design

The intention of the Chap. 6 was to demonstrate one of the core themes of this volume: that applied scientific questions themselves always imply questions about human social behavior. Observations regarding how emergency managers interact and use weather data, how forecasters negotiate weather risks, and the how the public responds served to address and inform several gaps in the development of DCAS. Yet, as was also seen, unraveling the assumptions of applied scientific questions, testing them, and building results into system design defies any straightforward or singular approach. Rather, interdisciplinary research involves the participation of multiple stakeholders, some of whom serve as scientific partners, while others take part as simply as subjects of investigation. One of the greatest challenges of interdisciplinary research was—and, arguably, remains—distinguishing genuine patterns of end-user thought and behavior relevant to system design from other instances of merely incidental behaviors, which, while impacting health and safety, cannot be integrated as their occurrence is too far removed from the system or too random to be governed through design. For instance, that emergency managers shared difficulties in interpreting velocity data. Since this forms a pattern, it could readily be incorporated into the design of ERC-CASA. On the other hand, there are also cases in which, by pure chance, members of the public failed to hear local sirens because they just so happened to be mowing the lawn at the time during sounding. Here there is no pattern, and, hence, no relation to system design. In other cases, poverty prevented people from purchasing adequate tornado shelters. While this indeed relates to tornadic vulnerability, resolving widespread poverty either directly or indirectly is far beyond the scope of CASA or any other technical problem; it is a broader societal issue. One of the most difficult challenges of interdisciplinary research moving ahead is to determine (a) which observations are part of patterns which can be incorporated into system design and (b) what observations are beyond the scope of the project or simply random noise for which the development of system policies is not realistic or practical.

Bibliography

Aguirre, B.E., D.E. Wenger, and G. Vigo. 1998. "A Test of the Emergent Norm Theory of Collective Behavior." Sociological Forum. 13: 301–320.

Associated Press. (2006). "Tornadoes Blow Through New Orleans." Accessed April 10, 2007.

Bass, E.J., Baumgart, L.A., Philips, B., Kloesel, K., Dougherty, K., Rodríguez, H., Díaz, W., Donner, W., Santos, J., and Zink, M. (2009). Incorporating Emergency Management Needs in the Development of Weather Radar Networks. *Journal of Emergency Management*. 7(1): 45–52.

Brotzge J.A., Droegemeier K., and McLaughlin D.J. (2006). Collaborative Adaptive Sensing of the Atmosphere: New Radar System for Improving Analysis and Forecasting of Surface Weather Conditions. *Transportation Research Record*, 2006, 1948(1), 144–151.

Brotzge, J., Hondl, K., Philips, B., Lemon, L., Bass, E. J., Rude, D., & Andra, D. L., Jr. (2010). Evaluation of Distributed Collaborative Adaptive Sensing for Detection of Low-Level Circulations and Implications for Severe Weather Warning Operations, *Weather and Forecasting*, *25*(1), 173–189.

Chandrasekar, V., D. McLaughlin, J. Brotzge, M. Zink, B. Philips, and Y. Wang, 2007: Distributed collaborative adaptive radar network: The CASA IP-1 network and tornado observations.

Preprints, 33rd Conf. on Radar Meteorology, Cairns, QLD, Australia, Amer. Meteor. Soc., 13A.3A. [Available online at http://ams.confex.com/ams/pdfpapers/128257.pdf.]

Chandrasekar, V., D. McLaughlin, J. Brotzge, M. Zink, B. Philips and Y. Wang. (2008). Distributed Collaborative Adaptive radar network: Preliminary results from the CASA IP1 testbed. *2008 IEEE Radar Conference*, 1–5, doi: https://doi.org/10.1109/RADAR.2008.4721125.

Donner, W. (2007). *An Integrated Model of Risk Perception and Protective Action*. Doctoral Dissertation, Disaster Research Center (DRC), University of Delaware, Newark, DE.

Donner, W., Rodriguez, H., and Diaz, W. (2012). "Tornado Warnings in Three Southern States: A Qualitative Analysis of Public Response Patterns." *Journal of Homeland Security and Emergency Management*. 9(2): 1–19.

Donner, W.R., Grainger, D., Rodriguez, H., Diaz, W., Santos, J. (2005). Rainfall Estimates or Tornado Detection?: An Assessment Based on the Needs of Emergency Managers. ERC-CASA End-User Integration, Research Brief No. 4, Disaster Research Center (DRC), University of Delaware, Newark, DE.

ERC-CASA. n.d-a. ERC CASA Overview. *Center for Collaborative Adaptive Sensing of the Atmosphere*. (http://www.casa.umass.edu/main/about_us/overview/).

ERC-CASA. n.d-b. End-user Integration. Accessed on 11/11/19 (http://www.casa.umass.edu/main/casansf/research/end_user_integration/)

Lai, J. and Widmar, N.O. (2021). Revisiting the Digital Divide in the COVID-19 Era. Applied Economic Perspectives and Policy, 43(1), 458–464.

League, C.E., Díaz, W., Philips, B., Bass, E., Kloesel, K., Gruntfest, E., and Gessner, A. (2010). Emergency Manager Decision-Making and Tornado Warning Communication. *Meteorological Applications*, 17(2), 163–72. doi: https://doi.org/10.1002/met.201.

Marks, D. 2006. Digital Inequality and the Implementation of New Technologies: Problems with Technological Diffusion among Oklahoma Emergency Managers. Preliminary Paper #334, Disaster Research Center (DRC), University of Delaware, Newark, DE.

McLaughlin, D. et al. (2009). Short-Wavelength Technology and the Potential For Distributed Networks of Small Radar Systems, *Bulletin of the American Meteorological Society*, *90*(12), 1797–1818.

Zink, M., Lyons, E., Westbrook, D., Kurose, J., and Pepyne, D. (2010). Closed-loop architecture for distributed collaborative adaptive sensing of the atmosphere: meteorological command and control. *International Journal of Sensor Networks*, 7(1–2), 4–18.

Mitchell, J.T., Thomas, D., Hill, A., and Cutter, S. (2000). Catastrophe in Reel Life versus Real Life: Perpetuating Disaster Myth through Hollywood Films. *International Journal of Mass Emergencies and Disasters*, 18(3), 383–402.

Philips, B. et al. (2007). Integrating end user needs into system design and operation: the Center for Collaborative Adaptive Sensing of the Atmosphere (CASA). 87th American Meteorological Society Annual Meeting, San Antonio, TX.

Red Cross. (2007). "Red Cross continues tornado assistance in New Orleans area." Accessed from the Internet April 10, 2007 (http://www.redcross.org/article/0,1072,0_312_6244,00.html).

Rodríguez, H., Donner, W., Díaz, W., and Hernández, J.S. (2010). "Emergency Managers, Allocation of Radar Resources, and Policy Implications: The Intersection of Weather Hazards, Population, and Technology." *Journal of Emergency Management,* 8(2), 35–44. doi: https://doi.org/10.5055/jem.2010.0010.

Rogers, Everett M. 2004. Diffusion of innovations Free Press: New York

Urban Institute. 2005. Katrina: Demographics of Disaster. Office of Public Affairs, Washington, D.C.

Chapter 8
Emerging Interdisciplinary Themes: Social Media

There has been a growth in research on the emerging role of social media in the mitigation, response to, and recovery from disaster (McCaffrey 2015). From roughly 2001 onwards, the field witnessed a growing number of studies focusing on the relationship between social media and disasters, the earliest of which being the Indian Ocean Tsunami (Liu et al. 2008), Hurricane Katrina (Endsley et al. 2014), 2011 active shooter in Norway (Perng et al. 2013), and more recently on the 2015 Nepal Earthquake (Soden and Palen 2016). Social media is used by emergency managers not only to communicate information but also build trust and legitimacy with the community audience (Panagiotopoulos et al. 2016). In a very short period of time, social media revolutionized how we communicate and the conditions under which communication occurs. It is perhaps an understatement to say that the internet, and, in particular, social media, have profoundly influenced the quality and quantity of mass communication in the past three decades. Since mass communication is central to warning response, and has therefore transformed risk communication in profound ways, it follows that interdisciplinary research will be required to take into account the evolving role of social media in the warning process. Where once telephone, televisions, and sirens were the principal means of communicating warnings, the internet, as a growing platform for risk communication, is becoming a dominant method—if not *the* dominant method—for both communicating and receiving severe weather warnings (either formally through institutions or informally through social networks). This evolution has held one key advantage: warnings can, at least in theory, be communicated more quickly to those at risk, and, all else equal, such improvements in the speed of communication should result in a net decrease in mortality, injuries, and property losses, largely due to increases in personal lead time. But this evaluation of the merits of social media may be far too reductive. To be sure, there are reasons to doubt that social media, in spite of the increased speed with which warnings are communicated, offers a clear net benefit to public safety as far as tornadoes are concerned. We simply do not yet know how, whether, and under what circumstances social media improves response.

W. Donner et al., *Severe Weather Warnings: An Interdisciplinary Approach*,
https://doi.org/10.1007/978-3-031-05031-2_8

How Does Social Media "Fit" into Our Current Understanding of Warnings?

Again, returning to the warning model is useful: how do warning communicated via the internet and social media impact reception? Does the public understand warnings better when received via social media? Are warnings received through Twitter, Facebook, and other platforms more credible and more likely to be personalized? Such questions are important, not only from the perspective of interdisciplinarity, but also the perspective of emergency management, given the increasing popularity of social media as a tool for public safety.

These changes hold natural implications for the study of warning systems from an interdisciplinary perspective. While research on the impact of social media on warning systems is, in relative terms, somewhat new, here we hope to provide a brief overview of some trends and patterns that may be relevant for the future academic pursuits of interdisciplinary warnings.

As should be apparent at this point in the discussion, interdisciplinarity acknowledges the need to adapt new technologies—such as new communications architectures—and social-systemic realities into the technological research enterprise. Thinking back to the warning model, social media, in changing fundamentally how we access information and how we communicate, is therefore all but assured to play an increasingly vital role in the processes by which warnings are received, evaluated, and responded to both at the institutional level (e.g., emergency management) and individual level (community). This naturally leads to several questions. Do platforms such as twitter and Facebook make weather information more or less *accessible* to the public, particularly for those who may find more traditional channels of communication inaccessible during periods of warning? Are warnings *better understood* through a social media interface, which is to ask whether social media discussions serve to illuminate or further mystify the meanings of warnings? Do social media platforms foster *skepticism* of warnings, or, alternatively, encourage the community to *believe* and take seriously severe weather threats? Such questions remain unanswered. In this chapter, we provide an overview the literature on social media and severe weather warnings, examining key trends in the use of social media in the response to and management of disasters. Notably, we address how public safety institutions use social media to alert the public to risk, as well as the role and consequences of social media use in public response to warnings.

Conceptualization of Social Media as a Warning Tool

Recent efforts to evaluate the role of social media in the warning and response phase of disaster have focused on several broad issues and themes. Sutton et al. (2015a, b), for instance, have examined Twitter and the role "terse" communication plays in the warning process. Another area of research—on the phenomena known as "crowdsourcing"—evaluates the value of mass observations and mass reporting of disaster events, addressing whether vast online communities detect and describe the scope, impact, and timeframe of hazards with enough accuracy and reliability for use as an emergency management tool. So too have the traditional sociological areas of inquiry in the context of social media and warning response been addressed, many of which examine the cultural, structural, and psychological dimensions of social media and warnings.

There is very little doubt that social media has fundamentally altered risk communication. Facebook, Twitter, and WhatsApp, in part due to the unprecedented social connectivity each offer, have altered the basic relationship between communities and disaster. Warnings, as a form of communication, have therefore been heavily affected in this regard. In the past, most warning information was communicated unidirectionally from agency to agency or from agency to community, but social media has allowed for more efficient and more robust community to community transmission of warning information (Kaewkitipong et al. 2012). While past warnings tended to circulate within primary groups, now discussion, debate, and interpretation of warning messages may more often take place among secondary groups on social media platforms. Accordingly, Kaewkitipong et al. suggest the following expansion of warning transmission that accounts for this new trend:

- Agency to agency
- Agency to community
- *Community to community*

Added to this model is yet another pattern of communication, however: *Community to agency* (via what was described earlier as "crowdsourcing"). Krennert et al. (2018: 75), for instance, notes an increasing willingness among European meteorologists to seek "non-conventional" means of weather information, wherein "formally or informally, nearly all forecasters from European NMHSs search for additional, non-conventional information about weather events and their impact. The ever-increasing facilitation of transmitting information, also including large data quantities like pictures and videos, in real-time fosters evidently rising importance of human ground impact observations." Reuter and Kaufhold (2018) and Reuter et al. (2012) argue that warning communication is no longer confined to transmission unidirectionally from authorities to citizens and authorities to authorities. Now, rather, online communication opens up the possibility of greater citizen-to-citizen communication and citizen-to-authority communication.

Indeed, the top-down, unidirectional model of warning communications may be less applicable nowadays as there appears to be a growing trend towards social-media-driven interaction between emergency managers and the public (Wukich and Mergel 2015). A problem with this development, however, is that what a "two-way" model of warning communication may look like in practice remains at present largely unknown (Lovari and Bowen 2020). In the following sections we discuss the consequences of these changes for the traditional warning process.

How Has Social Media Altered Traditional Disaster Processes?

Rumor

To better understand the role of social media in warning, it is important to understand the role of rumor. It is tempting to regard social media as a source of rampant, damaging rumors through which warnings are distorted, but, as with all things, the role of rumor is far more nuanced. As sociologists, we might first ask: what is the function of rumor generally. Rumor, a concept discussed widely in the collective behavior and social movement literature, is an important process in the context of interdisciplinary warnings because informal communication is critical to the establishment of belief in hazard risk, personalization, and confirmation. It is common to regard rumor as a process by which communication is distorted, but this is far too simple. Shibutani (1966), in one of the earliest and perhaps most instructive treatments of rumor, cautions us to view rumor less as simply false information and more as a social process that attempts to confront and resolve uncertainty. In other words, as uncertainty is ubiquitous, so too is the presence of rumor as the two exist in a functional relationship. This idea is incredibly important because much of the warning process involves a great deal of uncertainty, and therefore understanding rumor dynamics is critical, particularly as far as social media is concerned.

What, then, gives rise to rumor in the context of the warning process? Oh et al. (2013) attempted to develop a statistical model of rumor-generating variables, yielding several interesting results. Interestingly, it was found that it was generally not the *content ambiguity* (was the message vague?) of messages that resulted in rumor generation, but rather *source ambiguity* (was the source vague?). When the source of warning information was not well understood or trusted, rumors tended to emerge to resolve ambiguity created by this lack of confidence. This would seem to suggest that, as the literature on warnings has stated many times, trust in the organization from which information is received is critical to belief. Other factors have also been found to be predictive of rumor: information without a clear source, a degree of personal involvement, and anxiety tend to be drivers of the rumor processes during disaster (Oh et al. 2013). Chua et al. (2017) and Liu et al. (2014) et found similar findings, but, unlike Oh et al. (2013), also that content ambiguity indeed played a role in the rumor process. How might this impact the warning model? In terms of the stages, rumor may make the process of confirmation much more likely and may therefore additionally prolong the process of confirmation.

How might we combat rumor during the warning process? In a survey of Puerto Rican emergency managers, it was found that social media was used by authorities to effectively dispel rumors during a hazardous storm (Bui 2019). In other cases, there has been an observed tend among online communities to monitor and correct for false information, thereby potentially mitigating for the effect of rumor: "more encouragingly, the crowd was able to make substantial improvements to the raw message. Crowd members "voted up" relevant tweets at a substantially higher rate than they did other message types. . .relevant retweets were modal in the channel and dominated on occasion" (He et al. 2017: 41).

Crowdsourcing

Online crowdsourcing is an effort to obtain information from a large collection of people. But is such a tool useful? As far as risk communication is concerned, there appears to be a high degree of correlation between Twitter activity and the presence of watches and warnings:

> Tornado-related tweets are relatively frequent on days when a large number of tornado watches are issued (compared todays where few, if any, warnings or watches are issued), but even more frequent on days when a relatively large number of warnings are issued. A similar pattern characterizes the temporal relationship between Twitter activity and the number of people affected by warnings and/or watches—on days when a large number of people are included in watches, the number of tornado-related tweets is high (relative to the days where few, if any, people are affected), but it is even higher on days when a large number of people are impacted by tornado warnings. (Ripberger et al. 2014: 529)

Social media appears to follow patterns of warnings, but is social media activity useful in identifying the location of threats themselves? Evidence seems to suggest so. Earlier research reports that tweets may indeed be reliable and useful in spotting natural hazards. In one study, Twitter users were found to have identified earthquake activity in Japan with 96% accuracy (Sakaki et al. 2010). Similar outcomes were observed in other contexts through another study in which it was found that social media was similarly useful in predicting the location of heatwaves in Italy (Grasso et al. 2017), and later research found that a syntheses of forecast and social media activity performed exceptionally well at detecting ongoing disaster events (Rossi et al. 2018). Indeed, internet users may form what Grasso et al. (2017) call a "social radar" by which events are detected and located via crowdsourcing. According to findings, it would appear as if this "social radar" is more or less effective in general terms in detecting disasters.

With respect to severe weather specifically, however, the use of crowdsourcing as a predictive tool seems less straightforward. Halse et al. (2018: 7) straightforwardly report that "findings indicate that there was indeed a positive correlation between the amount of weather related posts and an increase in wind speed for a local area." Other research has painted a more complicated picture, however. Smith et al. (2015) examined Twitter observations of flood events in the United Kingdom to determine whether crowdsourcing efforts might provide an effective way to detect and predict severe weather. In this regard, the depth and scope of flooding events, according to the results, were well estimated and corresponded closely to the conclusions of weather models. On the other hand, social media was much less effective at identifying and determining rainfall amounts (perhaps because rainfall amounts are far more dynamic and unpredictable in comparison to floods). Also notable was the evolving function of social media as the floods progressed in time. In the beginning of a flood event, tweets focused heavily on the immediate shock or impact of flooding. As time passed, however, tweets were found to increasingly focus on the details and measurable dimensions of impact. Thus, crowdsourcing may be less useful as a tool in the beginning of the event and more useful as the event evolves. This might provide some measure of guidance if tweets are used to make decisions: emergency managers should wait for some time before using social media to determine the location and extent of severe weather.

The Psychological and Social Psychological Impacts of Social Media Warnings

Natural concerns have been raised over whether warnings communicated via social media invoke understanding, belief, and personalization in ways similar to more traditional systems of communication. Again, while social media is thought to—and most likely does—increase the visibility of warnings, the psychological and social psychological implications of internet-mediated risk communication remain nevertheless unclear. Again, viewed in the context of the warning response model, we see the matter of online risk communication to be far more complex than initially thought. In a recent focus group study, participants were asked to evaluate tweets intended to "warn" of an imaginary nuclear attack (Bean et al. 2016). The curt messaging characteristic of the tweet medium was found to be, according to the authors, "confusing, difficult to believe, and impersonal" (136). The authors summarize results accordingly:

> WEAs and tweets are a form of terse communication in that they limit the amount of message content, do not provide an opportunity for dialogue and require the message receiver to quickly take protective action. As theorised, terse communication can generate uncertainty, thereby prompting WEA and tweet recipients to 'mill' for additional and confirming information. (144)

Results seem to imply, as is consistent with the warning literature, that the abrupt nature of the Twitter format holds the effect of encouraging rather than resolving doubt, thereby prompting the need for confirmation. Thus, while Tweets may reach a larger number of people, response may be delayed as the structure of communication ("terse messages") unintentionally fosters the urge to search for additional, more detailed information. This line of reasoning appears to be supported in the literature as it is shown that shorter messages prompt confirmation behavior more often than longer messages (Wood et al. 2018), although later research would paint a more complicated picture of the issue (Sutton et al. 2018).

As well, both the cognitive and emotive features of warnings cannot be ignored. With respect to cognition, Wang et al. (2020) identified context features of a social media message that increased the likelihood of message retransmission during Hurricane Irma: information containing hazard, guidance, location, time, and source information were likely to be retweeted during the event. While tweets containing facts are important, the role of tweets that invoke strong feelings cannot be overstated, as messages invoking an emotional response are more likely to be shared and serially transmitted (Sutton et al. 2015a, b), an observation on which Gutteling et al. (2018: 1587) elaborate further: "Even though cognitive factors like assessed message quality are obvious prerequisites for behavior, our results show that emotions and social environment are main predictors for actually performing the required behavior." Emotions also seemed to furthermore enhance public response in the context of hazard mapping according to Liu et al. (2017). Tweets, therefore, must connect with both the rational and irrational elements of human psychology in order to minimize skepticism and the need for prolonged and unnecessary confirmation. The emotional dimension of tweets is perhaps not terribly surprising as they may be similar to "impact-based" warnings, the effects of which on response were discussed at length earlier.

Population and Culture

Apart from individual psychology, culture also shapes patterns of social media use among populations. Murthy and Longwell (2013), in an examination of Twitter use during the 2010 Pakistani floods, found that Western users were more likely to link traditional media sources. On the other hand, Pakistani twitter users were more likely to link social media sources, which perhaps suggest that local norms, values, and beliefs may shape trust in different news mediums. This notion is further supported in Kim (2011), who found that there were two significant differences between how Americans and Koreans used social media. First, Korean students seemed to use preexisting social networks for care and support whereas American typically used such networks for the purposes of entertainment. Second, Americans had larger online networks, whereas Koreans placed more emphasis on smaller, more intimate networks.

Also, the notion of "digital inequality" is relevant to the discussion of social media and warnings as the concept draws attention to concerns about gaps in technology access between social groups (Dimaggio et al. 2004; Rogers 2004). There is perhaps no greater gap in access than that between younger and older adults. According to a recent poll by the Pew Research Center (2017), senior citizens face significant barriers to the full adoption and use of modern technologies. According to the poll, while 74% of 18–29 years old and 60 % of 30–49 years felt "very confident" using electronic devices, by comparison far fewer seniors—roughly 26%—held comparable attitudes towards electronic devices. Thus, from an interdisciplinary perspective, even if Twitter were to be optimized in terms of a warning model, older adults may experience diminishing returns on such utility due to generational divide in social media use.

Conclusion

This chapter has been brief in the overview of literature on social media and warnings largely due to the presence of only a small body of literature on the matter. Nevertheless, there is great importance in acknowledging the role of social media in the warning process, as well as the relevance of social media for interdisciplinary warning research, as it has become the dominant medium for communicating both generally and in terms of warning specifically. Accordingly, there emerges fundamental changes in the speed with which information is communicated, rumor is spread, truth is evaluated, and populations may be reached and persuaded under new and unprecedented regimes of communication. Any new interdisciplinary warnings project must acknowledge this moving ahead. *The most important matter to consider is that at some point in the future social media will supplant traditional forms of media as a channel through which warnings are received and communicated.* What was fashionably called "new media" in the past has increasingly become the standard media. Accordingly, we must be prepared with an understanding of the new rules that govern the reception, understanding, and belief in warnings in an online environment if we are to fully adapt warning strategies to this new era of communication.

Bibliography

Anderson, M., and Perrin, A. (2017). Tech Adoption Climbs Among Older Americans. *Pew Research Center: Internet, Science & Tech*. (https://www.pewresearch.org/internet/2017/05/17/tech-adoption-climbs-among-older-adults/).

Bean, H., Liu, B.F., Madden, S., Sutton, J., Wood, M.M., and Mileti, D.S. (2016). Disaster Warnings in Your Pocket: How Audiences Interpret Mobile Alerts for an Unfamiliar Hazard. *Journal of Contingencies and Crisis Management*, 24(3):, 136–147. doi: https://doi.org/10.1111/1468-5973.12108.

Bhavaraju, T., Krishna, S., Beyney, C., and Nicholson C. (2019). Quantitative Analysis of Social Media Sensitivity to Natural Disasters. *International Journal of Disaster Risk Reduction*, 39, 101251. doi: https://doi.org/10.1016/j.ijdrr.2019.101251.

Bui, L. (2019). "Social Media, Rumors, and Hurricane Warning Systems in Puerto Rico." *Hawaii International Conference on System Sciences* 2019 (HICSS-52).

Chua, A., Tee, C., Pang, A., and Lim, E. (2017). The Retransmission of Rumor and Rumor Correction Messages on Twitter. *American Behavioral Scientist*, 61(7), 707–23. doi: https://doi.org/10.1177/0002764217717561.

Cox, J. and Plale, B. (2012). Improving Automatic Weather Observations with the Public Twitter Stream. Preliminary Manuscript.

Davis, E. (2018). How College Campuses Are Using Social Media During Severe Weather Events. *Graduate Theses and Dissertations* Retrieved from https://scholarworks.uark.edu/etd/2793

Dimaggio, P., Hargittai, E., Celeste, C., and Shafer, S. (2004). Digital Inequality: From Unequal Access to Differentiated Use. Russel Sage Foundation, 355–400.

Endsley, T., Wu, Y., Reep, J., Eep, J., & Reep, J. (2014). The source of the story: Evaluating the credibility of crisis information sources. In ISCRAM

Grasso, V., Crisci, A., Morabito, M., Nesi, P., and Pantaleo, G. (2017). Public crowdsensing of heat waves by social media data, Advances in Science and Research, 14, 217–226.

Gutteling, J.M., Terpstra, T. and Kerstholt, J.H. (2018). Citizens' Adaptive or Avoiding Behavioral Response to an Emergency Message on Their Mobile Phone. *Journal of Risk Research*, 21(12), 1579–91. doi: https://doi.org/10.1080/13669877.2017.1351477.

Halse, S., Montarnal, A., Tapia, A., Benaben, F. (2018). Bad Weather Coming: Linking Social Media and Weather Sensor Data. Pp. 507–515 in ISCRAM 2018 - 15th *International conference on Information Systems for Crisis Response and Management. Rochester,* United States: ISCRAM.

Hyvarinen, O., and Saltikoff, E. (2010). Social Media as a Source of Meteorological Observations. *Monthly Weather Review*, 138, 3175–84. doi: https://doi.org/10.1175/2010MWR3270.1.

He, X, Lu, D., Margolin, D., Wang, M., El Idrissi, S. and Lin, Y. (2017). The Signals and Noise: Actionable Information in Improvised Social Media Channels During a Disaster. In Proceedings of the 2017 ACM on Web Science Conference (WebSci '17). Association for Computing Machinery, New York, NY, USA, 33–42.

Kaewkitipong, L., Chen, C.C., and Ractham, P. (2012). Lessons Learned from the Use of Social Media in Combating a Crisis: A Case Study of 2011 Thailand Flooding Disaster. International Conference on Information Systems, ICIS 2012. 1. 766–782.

Kim, Y., Sohn, D., and Choi, S.M. (2011). Cultural Difference in Motivations for Using Social Network Sites: A Comparative Study of American and Korean College Students. *Computers in Human Behavior*, 27(1), 365–72. doi: https://doi.org/10.1016/j.chb.2010.08.015.

Krennert, T., Pistotnik, G., Kaltenberger, R., and Csekits, C. (2018). Crowdsourcing of Weather Observations at National Meteorological and Hydrological Services in Europe." *Advances in Science and Research*, 15, 71–76. doi: https://doi.org/10.5194/asr-15-71-2018.

Liu, S.B., Palen, L., Sutton, J., Hughes, A.L., and Vieweg, S. (2008). In Search of the Bigger Picture: The Emergent Role of OnLine Photo Sharing in Times of Disaster. Proceedings of the 5th International ISCRAM Conference – Washington, DC, USA, May 2008.

Liu, B., Wood, M., Egnoto, M., Bean, H., Sutton, J., Mileti, D., and Madden, S. (2017). Is a Picture Worth a Thousand Words? The Effects of Maps and Warning Messages on How Publics Respond to Disaster Information. *Public Relations Review*, 43(3), 493–506. doi: https://doi.org/10.1016/j.pubrev.2017.04.004.

Liu, F., Burton-Jones, A., and Xu, D. (2014). Rumors on social media in disasters: Extending transmission to retransmission. Proceedings - Pacific Asia Conference on Information Systems (PACIS) at AIS Electronic Library (AISeL) 17.

Lovari, A. and Bowen, S. 2020. "Social Media in Disaster Communication: A Case Study of Strategies, Barriers, and Ethical Implications." *Journal of Public Affairs*, 20, e1967. doi: https://doi.org/10.1002/pa.1967.

McCaffrey, S. (2015). Community Wildfire Preparedness: A Global State-of-the-Knowledge Summary of Social Science Research. *Current Forestry Reports*, 1(2), 81–90. doi: https://doi.org/10.1007/s40725-015-0015-7.

Mihunov, V., Lam, N., Zou, L., Wang, Z, and Wang, K. (2020). "Use of Twitter in Disaster Rescue: Lessons Learned from Hurricane Harvey." *International Journal of Digital Earth* 13(12):1454–66. doi: https://doi.org/10.1080/17538947.2020.1729879.

Misner, J.W. (2014). Mindful Leadership and Navigating the Seas of Change in the Information Age. *Journal of Leadership Studies*, 8(2), 46–50. doi: https://doi.org/10.1002/jls.21330.

Murthy, D. and Longwell, S.A. (2013). "Twitter and Disasters." *Information, Communication & Society,* 16(6), 837–55. doi: https://doi.org/10.1080/1369118X.2012.696123.

Oh, O., Agrawal, M., and Rao, H. (2013). Community Intelligence and Social Media Services: A Rumor Theoretic Analysis of Tweets During Social Crises. *MIS Quarterly, 37*(2), 407–426. doi: https://doi.org/10.25300/MISQ/2013/37.2.05.

Panagiotopoulos, P., Barnett, J., Bigdeli, A.Z., and Sams. S. (2016). Social Media in Emergency Management: Twitter as a Tool for Communicating Risks to the Public. *Technological Forecasting and Social Change*, 111, 86–96. doi: https://doi.org/10.1016/j.techfore.2016.06.010.

Perng, S., Büscher, M., Wood, L., Halvorsrud, R., Stiso, M., Ramirez, L., & Al-Akkad, A. (2013). Peripheral Response: Microblogging During the 22/7/2011 Norway Attacks. *International Journal of Information Systems for Crisis Response and Management (IJISCRAM), 5*(1), 41–57. http://doi.org/https://doi.org/10.4018/jiscrm.2013010103

Pew Research Center. (2017). Tech Adoption Climbs Among Older Adults. https://www.pewresearch.org/internet/2017/05/17/tech-adoption-climbs-among-older-adults/. Accessed 8/25/21.

Reuter, C and Kaufhold, M. (2018). Fifteen Years of Social Media in Emergencies: A Retrospective Review and Future Directions for Crisis Informatics. *Journal of Contingencies and Crisis Management*, 26(1), 41–57. doi: https://doi.org/10.1111/1468-5973.12196.

Reuter, C., Marx, A., & Pipek, V. (2012). Crisis management 2.0: Towards a systematization of social software use in crisis situations. International Journal of Information Systems for Crisis Response and Management (IJISCRAM), 4(1), 1-16.

Ripberger, J. T., Jenkins-Smith, H. C., Silva, C. L., Carlson, D. E., & Henderson, M. (2014). Social Media and Severe Weather: Do Tweets Provide a Valid Indicator of Public Attention to Severe Weather Risk Communication?, *Weather, Climate, and Society*, *6*(4), 520–530.

Rogers, Everett M. 2004. Diffusion of Innovations Free Press: New York.

Rossi, C. et al. (2018). Early detection and information extraction for weather-induced floods using social media streams. *International Journal of Disaster Risk Reduction*, 30(A), 145–157.

Sakaki, T., Okazaki, M., and Matsuo, Y. (2010). Earthquake Shakes Twitter Users: Real-Time Event Detection by Social Sensors. Pp. 851–60 *in Proceedings of the 19th international conference on World wide web, WWW '10. New York, NY, USA: Association for Computing Machinery.*

Shibutani, T. 1966. "Improvised News: A Sociological Study of Rumor". *Oxford, England*: Bobbs-Merrill.

Smith, L., Liang Q, James P, and Lin W. (2015). "Assessing the Utility of Social Media as a Data Source for Flood Risk Management Using a Real-time Modelling Framework" - *Journal of Flood Risk Management,* 10(3), 370–80.

Soden, R. and Palen, L. (2016). Infrastructure in the Wild: What Mapping in Post-Earthquake Nepal Reveals about Infrastructural Emergence. CHI '16: Proceedings of the 2016 CHI Conference on Human Factors in Computing Systems, 2796–2807.

Spence, P., Lachlan, K., Lin, X. and Greco, M. (2015). Variability in Twitter Content Across the Stages of a Natural Disaster: Implications for Crisis Communication. *Communication Quarterly* 63:171–86. doi: https://doi.org/10.1080/01463373.2015.1012219.

Starbird, K. and Palen, L. (2010). "Pass It On?: Retweeting in Mass Emergency." *Proceedings of the 7th International ISCRAM Conference* – Seattle, USA, May 2010 10.

Sutton, J., Gibson, C., Spiro, E., League, C., Fitzhugh, S., and Butts, C. 2015a. "What It Takes to Get Passed On: Message Content, Style, and Structure as Predictors of Retransmission in the Boston Marathon Bombing Response." *PloS One*. doi: https://doi.org/10.1371/journal.pone.0134452.

Sutton, J., Gibson, C., Phillips, N., Spiro, E., League, C., Johnson, B. Fitzhugh, S., and Butts, C. (2015b). A Cross-Hazard Analysis of Terse Message Retransmission on Twitter. *Proceedings of the National Academy of Sciences* 112 (48) 14793–14798. doi: https://doi.org/10.1073/pnas.1508916112.

Sutton, J., Vos, S. C., Wood, M. M., & Turner, M. (2018). Designing Effective Tsunami Messages: Examining the Role of Short Messages and Fear in Warning Response, *Weather, Climate, and Society*, *10*(1), 75–87.

Wang, W., Haase, T. and Yang, C. (2020). Warning Message Elements and Retweet Counts: An Analysis of Tweets Sent during Hurricane Irma. *Natural Hazards Review*, 21(1), n.p.

Wood MM, Mileti DS, Bean H, Liu BF, Sutton J, Madden S. Milling and Public Warnings. (2018). *Environment and Behavior*, 50(5), 535–566. doi:https://doi.org/10.1177/0013916517709561

Wukich, C. and Mergel, I. (2015). Closing the Citizen-Government Communication Gap: Content, Audience, and Network Analysis of Government Tweets. *Journal of Homeland Security and Emergency Management*, 12(3), 707–735.

Hu, X., Zhang, X., & Wei, J. (2019). Public Attention to Natural Hazard Warnings on Social Media in China, *Weather, Climate, and Society*, *11*(1), 183–197.

He, X, Lu, D., Margolin, D., Wang, M., El Idrissi, S. and Lin, Y. (2017). The Signals and Noise: Actionable Information in Improvised Social Media Channels During a Disaster. In Proceedings of the 2017 ACM on Web Science Conference (WebSci '17). Association for Computing Machinery, New York, NY, USA, 33–42.

Chapter 9
Conclusion

We conclude the book by briefly returning to the prologue, recalling an NSF quote of central relevance. Notably: "[an engineered system is so defined] as a combination of components that work in synergy to collectively perform a useful function." The definition is important. It suggests—strongly so—that technical innovation should not be regarded as an end in itself, but rather a means to an end, culminating in a "useful function" through the meaningful integration of human thought and behavior into system design. For, as has been argued throughout, technologies in a state of either development or refinement may or may not be useful to people—that is, may or may not be of benefit to health, livelihood, or security. Therefore, interdisciplinarity should not be regarded as simply an adjunct to engineering or applied research questions generally, but as a component of equal weight in the answering of applied scientific questions, without which the "useful function" to which such efforts are ultimately directed would be unachievable. With respect to severe weather, the useful function of any technology or model should therefore be to protect lives and property from tornadoes, floods, and hurricanes. To regard the aim in any other way misses the point of interdisciplinarity.

Where Do We Go from Here? Policy Towards Addressing Unanswered Questions and Future Directions

Whether regarding the weather, warnings, or other topic, the future of interdisciplinary research rests with several unanswered questions the resolution of which is required in order to make interdisciplinarity a common feature of applied scientific research. The following questions are central to this effort:

1. **How do we provide sufficient attention and resources to social science research while, at the same time, ensuring that the natural science elements receive enough funding to develop technology in a manner had the social**

W. Donner et al., *Severe Weather Warnings: An Interdisciplinary Approach*,
https://doi.org/10.1007/978-3-031-05031-2_9

scientists not been present on the project. Again, this is key: the integration of social science should enhance, and not impede, inquiry into storm structure, remote sensing, or any other theory or technology for which the project was developed. Nor, on the other hand, should the activities of natural scientists and engineers compromise the work of social scientists associated with the interdisciplinary project. As, once again, the overall mandate of an interdisciplinary project is to resolve a "global question" (e.g., reduce tornado or hurricane mortality), if funding or resources are lacking to fully achieve either, or integrate the two, then the project will likely evolve into a series of independent questions without much focus on the global question. An interdisciplinary plan from the outset, in collaboration with mentors and facilitators, builds a global question around which the budget will be built. Once the global question is in place, and collaborative subprojects developed, which will involve deriving testable assumptions and methodology, only then can a realistic budget that fully funds the anticipated activities of both the natural and social scientists be developed. Ideally, this should occur prior to the development of a grant proposal prior to submission. Naturally, this will require a bit more planning, organization, and communication, but this is the nature of interdisciplinary research that simply must be accommodated. All else equal, interdisciplinary research will simply take more time to develop. It stands that the goal is to produce the best possible technology guided by social scientific data and reasoning.

2. **According to the plans set forth in this volume, we will need trained and committed interdisciplinary scholars charged with integrating the social and natural sciences**. Identifying and recruiting interdisciplinary leadership (i.e., mentors and facilitators, discussed earlier) into the initial development of an interdisciplinary research project is crucial to the long-term success of an interdisciplinary program. Once again, the inclusion of an interdisciplinary specialist ensures the subprojects remain well integrated, which is to say that all research questions are channeled towards the global question the project seeks to address. Doing so maintains a collaborative focus, avoiding an ad hoc approach towards interdisciplinary research development, instead establishing an interdisciplinary vision as a core inaugural feature of a project and fostering partnerships among scholars from backgrounds between which collaboration may have been previously limited. Many scholars and researchers simply do not know how to develop interdisciplinary research or work collaboratively towards a global question. The role would be filled by a person preferably with some background or training in interdisciplinary research, as well as a familiarity with both the theories and methodologies of social and natural science. At present, no such training program exists to meet the demand for trained interdisciplinary mentors or facilitators on interdisciplinary projects. Moving forward, a way to achieve the vision might be through the founding and development of advanced degrees (MA degrees) with specializations in scientific interdisciplinarity. In order to do so, a consortium of social and natural scientists must come together, develop a general vision for the program, and assemble a philosophy and curriculum.

3. **Develop a model for identifying the social science assumptions of natural science questions**. Once again, a core thesis of this monograph is that interdisciplinary research regards applied natural scientific questions as containing implicit assumptions about human behavior. In order to maximize the benefit of the question for society, these assumptions must be tested so that the technology aligns—or, in the language of interdisciplinarity, integrates—with the behavior, thoughts, and needs among those to whom the technology is of a proposed benefit. Chapters 5, 6, and 7 provide some guidance in this regard, but there must be work specifically devoted to developing a general framework by which applied scientific questions can be examined for behavioral assumptions and those assumptions can be transformed into testable hypotheses.
4. **Identify and refine an ideal organizational structure through which to organize the various activities, roles, and responsibilities of an interdisciplinary project**. Chapter 5 dealt with the matter in some depth, but scholars should devote future attention to developing the model in more depth and detail. The mentor and facilitator within an interdisciplinary project would hold key roles in the development of the interdisciplinary character of the project. The mentor, once again, would work from the beginning to establish an interdisciplinary culture of collaboration, work collaboratively with natural and social scientists to establish a global question, and further work to ensure that subprojects work towards fulfilling the aims of that global question. The facilitator, once again, performs a more "nuts-and-bolts" function, offering feedback and advice on individual subprojects. At the very least, there should be a role that is dedicated to building an maintaining an interdisciplinary culture across the life of a project.
5. **We believe one of the more important lessons of this book is approaching the global questions as a series of applied scientific questions**. The foundation for an interdisciplinary project is the global question and a focus on the global question must be maintained. If a focus on the global question is lost, then, once again, the project risks breaking apart into a collection of decentralized groups. The engineers work discretely on one project; the seismologists on another; and the social scientists on another. Indeed, here there may be innovation, in a sense, but such innovation may lack integration with the social system—or, per the previous definition, component—and may therefore be of little benefit to communities. A focus on the global question, where, once again, project results are articulated in relation to some overall social benefit, at the very least serves as a calibration tool or benchmark against which the collaborative efforts of a project may be measured and compared. This is not to say that the establishment of a global question will solve all problems, but it at the very least creates a set of expectations for all members.
6. **Distinguish between patterns and non-patterns**. When integrating data from the social and behavior science into the system design process, one must ask: "what findings may be integrated into the system and what findings, though important to the 'grand question' remain beyond the general reach and scope of

the project." As noted earlier, an understanding of how lead time affects public safety relates directly to system design. If hours and days of lead time make communities less safe, as hypothetically noted, then there is no point in developing a system the intention of which is to extend lead times to similar lengths. In this way, an understanding of human behavior is fed back into system design, and the design approach is altered to accommodate this reality. In other words, the assumptions, being false, were corrected. On the other hand, there are random behaviors or patterns of behavior, though relevant to tornado safety, that may not be able to directly inform system design processes. For instance, in one of the CASA surveys, a respondent reported missing a warning because he was mowing the lawn. While relevant to tornado safety, this mere instance in someone failing to respond is of a far too random nature to sensibly inform system design.

A Final Note on Complexity, Risk, and Interdisciplinary Research

Even systems that are thought to be simple are deeply complex. The "simple" blare of a siren, the "unambiguous" flashing of an alarm, the "direct" language on a cigarette package betray a deeply complex and complicated process of perception and behavior. While this book has dealt with warnings as a principal subject, there is perhaps a broader lesson about where science and society are heading as systems and structures evolve to become more vast, complicated, and interconnected, particularly within the context of disasters. Accelerated trends towards social complexity, if left unchecked and unmanaged, yield collateral risks of a scope and character unprecedented in historical record. This is because complex systems conceal the origins and muddy the prediction of future risks. Cause and effect are difficult to determine in complex systems (Mayr 1961). So, it is thought, as complexity continues to grow, as a consequence we shall inevitably see a rapid and manifold increase in the risks we face on a day to day basis (Beck and Ritter 1992).

It will become increasingly difficult to determine how people fit into these complex systems, and, accordingly, how the *useful function* will be fully derived from the system. Indeed, we may know what the useful function is, what the system is intended to do to help people, but it will become harder and harder to get there. While the integration of power supplies with the internet, for instance, has solved several problems, it has opened up the U.S. power supply to increasing attacks from state and non-state actors. Thus, this has increased the complexity of power supply management, which has in turn increased the risks we face because new human systems have been introduced into a system that had in the past been of far simpler nature. Our growing incursion into coastal environments has led to a host of unprecedented "natural" disasters (when, in reality, all disasters are by definition "human made," most sociologists of disasters would assertively submit). The "useful functions" that engineering projects attempt to bring about in these realms are increasingly difficult to realize because there are so many different factors, so many different variables, and so many sources of uncertainty at play. Now, time is of the essence, since as society becomes more complex, the processes and methodologies through which we arrive at a "useful function" will also become just as complicated. This is why there is a pressing need to better articulate the theory of interdisciplinarity as well as articulate the ideal organizational form through which interdisciplinary research might be most successful.

Here it is important to highlight several key points made in the previous chapters.

- Warning response on the part of the public is not merely a single act, but rather a process by which warning response is undertaken. To regard warning response as a single act isolated in context risks the development of theory and policy misaligned with the reality of how people behave during severe weather.
- Neither an exclusive natural or social scientific perspective is sufficient to answer questions of growing complexity.
- Human and natural systems do not merely interact at one level, but at many levels, adding further complexity to what is rightly regarded as already highly complex.
- To build a successful, long-term interdisciplinary project, an organizational structure is necessary that includes both natural and social scientists, but also includes specialized roles whose task it is to provide guidance on how to best synthesize findings.

In closing, as was seen in the previous sections, it can hardly be argued otherwise that factors beyond the technology at hand shape and govern how the public responds to severe weather warnings and how institutions—particularly, forecasters and emergency managers—make the ultimate decision to issue a warning, neither of which are simple matters, involving complex processes in which different kinds of information are acquired, scrutinized, and applied in a variety of different contexts. Severe weather response should be thought of as a chain. On one end, there are the scientists who develop warnings and radar technologies so as to provide faster and more accurate information. Down the line are the forecasters and emergency managers, who interpret and use this information to either warn or not warn the public. Finally, at the end of the chain are the beneficiary end-users and stakeholders—the public, businesses, schools—who use this information in the final act of this response process. And, as the saying goes: a chain is only as strong as its weakest link. If the state of warning technology is poor, then neither emergency managers nor the pubic, however motivated to respond, will be able to respond because they will not receive warnings in a timely manner. If a well designed radar system—or any system, for that matter—imposes upon forecasters too much information, the warning decision making process will be compromised, and the public may not receive warning information until it is too late. If scientists develop technology that can detect and warn tornadoes with great precision, and emergency managers use this information to inform the public in a timely manner, if the public does not respond, or responds inappropriately, then advanced technology and strong decision making among emergency managers will mean little. Closing the metaphor, interdisciplinary research ensures that each link in the chain—from the scientists to the emergency managers and forecasters to the general public—is strong and fully connected to one another. We hope we have provided in this book some guidance in making the strongest chain as possible.

Bibliography

Beck, U., and Ritter, M. (1992). *Risk society: Towards a new modernity*. London: Sage Publications.

Mayr, E. (1961). Cause and Effect in Biology. *Science*, Vol. 134, Issue 3489, pp. 1501–1506.

The manufacturer's authorised representative in the EU is Springer Nature Customer Service Centre GmbH, Europaplatz 3, 69115 Heidelberg, Germany. If you have any concerns regarding our products, please contact ProductSafety@springernature.com

Printed and bound by CPI Group (UK) Ltd, Croydon, CR0 4YY
01/10/2026
02225893-0005